SPACE AND THE 'MARCH OF MIND'

Space and the 'March of Mind'

Literature and the Physical Sciences in Britain, 1815–1850

ALICE JENKINS

OXFORD
UNIVERSITY PRESS

OXFORD

UNIVERSITY PRESS

Great Clarendon Street, Oxford OX2 6DP

Oxford University Press is a department of the University of Oxford.
It furthers the University's objective of excellence in research, scholarship,
and education by publishing worldwide in

Oxford New York

Auckland Cape Town Dar es Salaam Hong Kong Karachi
Kuala Lumpur Madrid Melbourne Mexico City Nairobi
New Delhi Shanghai Taipei Toronto

With offices in

Argentina Austria Brazil Chile Czech Republic France Greece
Guatemala Hungary Italy Japan Poland Portugal Singapore
South Korea Switzerland Thailand Turkey Ukraine Vietnam

Oxford is a registered trade mark of Oxford University Press
in the UK and in certain other countries

Published in the United States
by Oxford University Press Inc., New York

British Library Cataloguing in Publication Data

Data available

Library of Congress Cataloging in Publication Data

Data available

Typeset by Laserwords Private Limited, Chennai, India
Printed in Great Britain
on acid-free paper by
Biddles Ltd., King's Lynn, Norfolk

ISBN 978–0–19–920992–7

3 5 7 9 10 8 6 4 2

For my parents, Phyl and H. J.

Acknowledgements

Several extraordinarily kind friends have read substantial portions of the drafts of this book. I do not think that the book would have been written without Juliet John, who read almost all of it in several different versions and often saw my point before I did. Susan Castillo was a tremendously generous good cop. Catherine Steel was a brilliantly effective bad cop and also gave me unstinting encouragement and solace, for which I am deeply grateful.

Colleagues and former colleagues at Glasgow University have been very generous with time and information: Alex Benchimol, Kirstie Blair, Vicky Gunn, Vassiliki Kolocotroni, Donald Mackenzie, Iain McClure, Costas Panayotakis, Seamus Perry, Andrew Radford, Nick Selby, Alex Thomson, Duncan Wu, and Alec Yearling. Susan Castillo and Nicola Trott ran the English Literature Department at Glasgow during the years in which I was writing this book and were tirelessly encouraging. Norman Gray, Nigel Leaok, Farah Mendlesohn, Stuart Robertson, Sean Sollé, Stephen Swann, Susan Stuart, and Nicky Trott again kindly read parts of the book and discussed key ideas with me. Frank A. J. L. James, Simon Schaffer, and Alan Rauch in particular have been especially generous with advice and information. I am very grateful indeed to my brother, Geoff Jenkins, who gave me a great deal of helpful advice throughout the process of writing this book.

The help of staff of Glasgow University Library and its Special Collections Library, as well as of staff in the Rare Books Rooms of the British Library and Cambridge University Library, was essential to the research for this project. I also want to thank the archives staff at the Royal Institution and Institute of Electrical Engineers.

I am very grateful to Andrew McNeillie at OUP for his enthusiasm for this book, to two anonymous readers for their helpful comments, and to Tom Perridge, Elizabeth Robottom and Jacqueline Baker for bringing the book to production.

Like many Victorianists and literature and science scholars of my generation, I owe an enormous debt to Gillian Beer for her enabling example both on the page and in the classroom.

This book is dedicated with love to my parents, in gratitude for thirty-six years of support and encouragement.

Contents

Introduction

'Space' is very much on the agenda these days.[1]

This book is an interdisciplinary study of British literary and scientific culture in the first half of the nineteenth century, focusing on the development in those decades of a new spatial imagination. Its central concern reflects the increasing importance of space in critical enquiry over recent decades. Since the mid-1980s, large areas of literary scholarship have begun to take the 'geographic turn'. Space and place in and around texts are now routinely topics of critical inquiry. Geography's basic conceptual toolkit—'specificity and spatiality, territoriality and locality, site and situation', in David N. Livingstone's words—is being adopted and adapted to serve in an enormously wide range of literary critical contexts.[2] This toolkit has obvious applicability to critical arenas such as postcolonial studies, where geography is one of the explicit conditions shaping the production and reception of texts. Equally, though, the enormous influence of Foucault on the critical academy has broadened the dissemination of geographical concerns: even more than the work of the great Marxist theorist Henri Lefebvre, Foucault's work has shown literary studies how space and place provide models for exploring the effects of power within systems, textual or

---

[1] Doreen Massey, *Space, Place and Gender* (Cambridge: Polity, 1994), p. 249.
[2] David N. Livingstone, 'The Spaces of Knowledge: Contributions Towards a Historical Geography of Science', *Environment and Planning D: Society and Space*, 13 (1995), 5–34 (p. 5). Of course the traffic between English studies and geography is two-way: as the distinguished critical geographer Yi-Fu Tuan pointed out in 1991, 'the increasingly common application of the word 'text' to architecture and landscape suggests, in itself, how deeply the linguistic viewpoint has penetrated the geographer's world' ('Language and the Making of Place: A Narrative-Descriptive Approach', *Annals of the Association of American Geographers*, 81 (1991), 684–96 (p. 685).)

other, thus contributing to the geographic turn in feminist and Marxist criticism.[3]

Until fairly recently space and time as objects of critical inquiry were to some extent polarized: attention to one seemed to preclude attention to the other. It is an important part of the achievement of the geographic turn in literary studies that this polarization seems in large part to have disappeared. In the mid-1980s David Harvey complained that many of the major modern social theorists devalued space and spatial concerns: 'they prioritise time and history over space and geography and, where they treat of the latter at all, tend to view them unproblematically as the stable context or site for historical action'.[4] More colourfully, Foucault described having been 'firebombed' by a Sartrean psychologist who claimed that '*space* is reactionary and capitalist, but *history* and *becoming* are revolutionary'.[5] Since Kant, Foucault argued, philosophy had devalued space, which was thought to stand 'on the side of the understanding, the analytical, the conceptual, the dead, the fixed, the inert'.[6] Any such valorization of interest in time at the expense of interest in space is all but absurd in the context of literary studies. Literary studies works with the medium of written verbal language, where—at the most basic level, the level of decoding marks on a page—all utterances necessarily depend equally on time and space for meaning to be produced.

But at a more sophisticated level of analysis, the general tendency of literary studies' geographic turn has been to reject the dogma that space is somehow unpolitical and unresponsive to ideology. This is partly because one of the driving factors in bringing geographic concerns into literary criticism has been identity politics, via feminist and postcolonial theory, which has clear commitments to political engagement with the world beyond criticism. The resulting rediscovery of the politics of

---

[3] Foucault's impact on literary criticism and the geographic turn is most clearly visible in the widespread take-up of his work on the panopticon and on heterotopia: central essays are 'The Eye of Power', in *Power/Knowledge: Selected Interviews and Other Writings, 1972–1977*, ed. Colin Gordon, trans. Colin Gordon *et al.* (Brighton: Harvester, 1980), pp. 146–65, 'Of Other Spaces', *Diacritics*, 16 (1986), 22–7, and 'Questions on Geography', in *Power/Knowledge*, pp. 63–77. Henri Lefebvre, *The Production of Space*, trans. Donald Nicholson-Smith (Oxford: Basil Blackwell, 1991).

[4] David Harvey, 'The Geopolitics of Capitalism', in *Social Relations and Spatial Structures*, ed. Derek Gregory and John Urry (Basingstoke: Macmillan, 1985), pp. 128–63 (p. 141).

[5] Michel Foucault, 'Space, Knowledge and Power', in *The Foucault Reader*, ed. Paul Rabinow (Harmondsworth: Penguin, 1984), p. 252.

[6] Foucault, 'The Eye of Power', pp. 149–50.

space in representation has become widespread through geographically-inclined literary criticism—so much so that I would argue that we are in danger of becoming blind to some of the inflections of the spatial imagination as it worked in times whose political and intellectual structures were different from our own.

One of the aims of this book is to develop a more accurate and wide-ranging understanding of the politics of space in the early nineteenth century, one that acknowledges that for some writers of the period space could indeed represent an ideal realm free from the taint of political concerns and human identity. And to achieve this better-calibrated understanding I am taking a step back from a focus on political representations of geographical space and instead investigating what I call the spatial imagination, a move which to some extent bucks the trend in nineteenth-century literary studies.

Nineteenth-century studies have been strongly responsive to the awakening of interest among literary critics in the possibilities of a critical investigation of space. Franco Moretti's *Atlas of the European Novel* stands as testimony to the originating and synthesizing power of geographical work on the period's literature, and though many of the suggestions made by Moretti have not yet been taken up in detail, a considerable body of scholarship has been produced on nations, houses, cities, streets, institutions, journeys, travel, and other geographical objects and activities in nineteenth-century literature of all genres.[7] Robert L. Patten's-stimulating essay on what he calls 'narrative traversals' in Victorian fiction sums up the concerns of this kind of criticism in a series of questions which he poses to his own work, but which can be read more widely as constituting some of the central questions asked by critics interested in 'space' in nineteenth-century literature:

---

[7] A selection of recent examples in this corpus: Thad Logan, *The Victorian Parlour* (Cambridge: Cambridge University Press, 2001); Sharon Marcus, *Apartment Stories: City and Home in Nineteenth-Century Paris and London* (Berkeley: University of California Press, 1999); Judith R. Walkowitz, *City of Dreadful Delight: Narratives of Sexual Danger in Late-Victorian London* (London: Virago, 1992); Deborah Epstein Nord, *Walking the Victorian Streets: Women, Representation, and the City* (Ithaca, N.Y.: Cornell University Press, 1995); Lynda Nead, 'Mapping the Self: Gender, Space and Modernity in Mid-Victorian London', in Roy Porter, ed., *Rewriting the Self: Histories from the Renaissance to the Present* (London: Routledge, 1997), pp. 167–85; Katherine Newey, 'Attic Windows and Street Scenes: Victorian Images of the City on the Stage', *Victorian Literature and Culture*, 25.2 (1997), 253–62; David L. Pike, 'Underground Theater: Subterranean Spaces on the London Stage', *Nineteenth Century Studies*, 13 (1999), 102–38.

Are there some ways that British nineteenth-century literature organizes the geographic places of its narratives? At different times in the century, do different kinds of spaces seem to predominate in literary representations? Do these different spaces incarnate particular sets of values and 'belong' to particular kinds of people? Does movement to, into, and away from these spaces signify differently not just in different works but more generally in different periods or sites? And might there be suggestive connections between particular spaces within texts and the formats of those texts, between particular kinds of travel to and from those spaces and specific modes of publication?[8]

Patten's list of questions bears witness to the combined historical, formal, and ideological inquiries raised by giving detailed attention to the geography of literary narratives. But beyond the local and geographical there is a different kind of space, one which does not boil down to a collection of places, but which certainly has the 'suggestive connections' with genres and modes of publication that Patten rightly proposes for narrative's geographies. So if it does not consist of particular places, what is this kind of space?

The kind of space discussed in this book is neither the lived space of social practice (rooms, streets, cities), nor the described space of geography (nations, continents, maps). Instead, it is the immaterial, conceptual space that contains and informs those other kinds of space; the space that allows us to perceive and compare distances, sizes, and locations. This is the kind of space, my book argues, that allows us to think about *things* at all, and that allowed the early nineteenth century to think about how to comprehend and organize its *things* at a moment when formerly stable principles of organization were in jeopardy.

In order to be transmitted and received, any kind of information has to be arranged. Arrangement in turn relies on determining relations between one element in a system and some or all of the others, and this applies whether the system is a drawerful of socks, an encyclopaedia full of facts or a novel full of characters. The kinds of relations one element can have with another in the system are enormously numerous, but some important ones include relations of similarity and dissimilarity, inclusion and exclusion, priority and anteriority. These kinds of relation are, and were in the early nineteenth century, often conceptualized in a semi-metaphorical way as spatial, because without a basic sense of space

[8] Robert L. Patten, 'From House to Square to Street: Narrative Traversals', in Helena Michie and Ronald R. Thomas, eds., *Nineteenth-Century Geographies: The Transformation of Space from the Victorian Age to the American Century* (New Brunswick, N.J.: Rutgers University Press, 2003), pp. 91–206 (p. 191).

we would not be able to understand whether one thing is inside another, whether it is bigger than another, whether it touches another. The same kinds of basic relationship that allow us to organize a sock drawer also allow us to organize sets of information about things that we cannot see and hold, such as our knowledge of the world beyond the range of our immediate sense experience.

Or, to take an early nineteenth-century example, the same kinds of mental processes that allow us to perceive the organization of a landscape are analogous to the ones that allow us to perceive the organization of a body of knowledge. Whether this claim is true according to the tenets of modern neuropsychology or even nineteenth-century philosophy of mind is not important to my argument. What matters is that the claim (and scores of others like it) was made, over and over again, in all manner of different contexts and to different audiences.

This book, then, steps back from the lived spaces of social geography to view the abstract space with and within which early nineteenth century writers organized knowledge, and which affected literature and the literary imagination in ways which have not yet been fully investigated by literary critics. In doing so, this book's project runs the risk of falling into what Henri Lefebvre called a 'basic sophistry'. Lefebvre claimed that a philosophical school he associated with Derrida, Barthes and Kristeva had had a politically deleterious effect on the study of space. It had 'fetishized' imagined or conceptual space and ignored real, lived spaces, such that 'the mental realm [came] to envelop the social and physical ones'. The result was an analysis of space in which 'a powerful ideological tendency [...] is expressing, in an admirably unconscious manner, those dominant ideas which are perforce the ideas of the dominant class'.[9] Instead of this emphasis on mental space, Lefebvre insisted on a profound engagement with the sociopolitical geography of ordinary human existence within physical environments.

It is certainly true that the kind of mental, or conceptual, or abstract space focused on in this book exists in a dynamic relationship with the lived geography of everyday life and thus is neither universal nor timeless, but rather is subject to variation and change as material conditions differ and alter. And this is where the bulk of interest in space in nineteenth-century studies has concentrated: on the lived geography of the characters within the text and—to a rather lesser extent—on the

---

[9] Henri Lefebvre, *The Production of Space*, trans. Donald Nicholson-Smith (Oxford: Basil Blackwell, 1991), pp. 5–6.

effect of lived geography on the production and distribution of texts. But the fact remains that a great deal of imaginative energy in the early nineteenth century went into describing abstract space as so far beyond the human and the historical that it could be relied on when the here and now failed. Where literary criticism has dealt with this kind of space, it has generally done so by labelling it as 'transcendent' or 'sublime', often with the effect of opposing it to the social and political and limiting it to the realm of the aesthetic or the psychological.

If we want to understand the early nineteenth century's habits of thought, particularly its ways of coping with its own burgeoning knowledges, we must understand its ideas about space. Those ideas were rooted in a large number of what we would now think of as diverse disciplinary areas, probably most importantly philosophy and the physical sciences (broadly speaking, chemistry and physics). And they were expressed in extraordinarily various genres and public forms, from the novel and poetry through political and journalistic writing to textbooks and treatises. It would be perfectly possible to base a study of the idea of space in early nineteenth-century British culture on the evidence of almost any grouping of these kinds of text. This book does not attempt to provide or even suggest a synthesis of all these possible studies. Instead, following the lead set by Gillian Beer, Sally Shuttleworth, George Levine, and others, I have used the interrelations of scientific and literary culture in the period as the main filter through which to make sense of space and to understand contemporary writers as they themselves tried to make sense of it.

It might seem counterintuitive to use the physical sciences with their focus on the material world as the sources for a book concentrating on immaterial space. Would not these sciences be better suited to supporting an argument about physical space? The correlation is not nearly so clearcut as that question assumes. One of the debates highlighted in this book, particularly in Part II, is the early nineteenth-century argument about science and materialism. Did physics and chemistry show that the universe could be explained without recourse to immaterial agencies, crucially that of God, or on the contrary were the hard sciences increasingly describing the universe as immaterial and undermining the definition of matter? Arguments on the latter side of the question were stronger both in quality and quantity than simplistic accounts of early nineteenth-century science acknowledge. To equate the physical sciences in this period with an unambiguously materialist outlook is gravely to misread the evidence.

Massively expanding literacy and print culture in the early nineteenth century resulted in unprecedented public demand for unprecedented amounts of new knowledge. This book's fundamental interest is in contemporary strategies for dealing with the disorder produced by this expansion and politicization of what Coleridge named 'the reading public'. Alan Rauch rightly argues that in the nineteenth century 'direct access to knowledge, through popular, cheap, and readable texts, became a central factor in both the production of knowledge and the structuring of social order'.[10] The impact of the reading public's drives to knowledge on social order and disorder has been traced by historians including E. P. Thompson and Richard Johnson; but we have much less information about the use of intellectual and epistemological structures in managing the changes produced by the new knowledge economy.[11] This book proposes that because many ways of managing and organizing systems (including systems of information) depend on spatial thinking, a focus on early nineteenth-century orderings of information demands attention to the contemporary spatial imagination.

Exploring contemporary conceptualizations of space allows us to perceive some of the strategies used to regulate access to knowledge and to manage the twin pulls towards spread and containment of information. Some of these strategies involved use of spatial metaphor, for instance by imagining knowledge as a landscape in which certain kinds of journeys and certain kinds of traveller were permitted and others excluded. Other strategies included organizing information by means of hierarchies, disciplines, and other territorial entities, which depended on spatial models for their logical and rhetorical force. Closely tied up with these spatial metaphors and models were a series of cultural responses to physical, geographical space—for instance, trends in horticulture and landscaping, and attitudes to the property-holding system or to technologies of travel. In the transfer from discourses directly referring to physical space to discourses in which space is metaphysical, conceptual, or imaginary, these cultural responses were not lost but were subject to transformation.

The physical sciences work in two ways in this model. For one thing, they provided subject matter, generating new information for which

---

[10] Alan Rauch, *Useful Knowledge: The Victorians, Morality, and the March of Intellect* (Durham, N.C.: Duke University Press, 2001), p. 14.

[11] E. P. Thompson, *The Making of the English Working Class* (London: Gollancz, 1963); Richard Johnson, 'Educational Policy and Social Control in Early Victorian England', *Past And Present* 49 (November 1970), 96–119.

the public had an appetite and which had to be managed. At the same time they contributed to changes in basic concepts of space, which in turn affected ways of organizing and disciplining information. The same could be said of other sciences, particularly those like geology which deal directly with the physical environment. Following Gillian Beer's seminal *Darwin's Plots*, the burgeoning academic field of literature and science studies has certainly paid more attention to literary culture's relationship with the life sciences and earth sciences than to its interactions with physics and chemistry. The impact on Victorian culture of geology, for example, has received a good deal of interdisciplinary critical attention over recent years.[12] But where these studies touch on matters of space, the space discussed is generally material and physical rather than conceptual.

It is through the interactions of nineteenth-century physics and chemistry with broader contemporary culture that we can best perceive the ways in which science contributed to developing ideas about immaterial space. New concepts of the nature of space and the matter that exists within it were being articulated in arguments about the forces holding together atoms and molecules and about the nature of the transmission of forces such as light, electricity, and magnetism. If we want to trace the development of a new understanding of space in the first half of the nineteenth century, we need to look to contemporary work in physics and chemistry. And we need to look to those disciplines' relationships with literary culture if we want to see how different concepts of 'space' were traded between and among those discourses in the emergent public sphere of the period—a public sphere which relied on spatial tropes to structure and regulate its own existence.

The emerging public sphere provides several of the reasons for focusing a study of space, literature, and the physical sciences on the first half of the nineteenth century. Despite its political and social fractures, this historical moment was also one of extraordinary cultural commonality among the educated reading public. It has often been noted, for example, that periodicals in the early nineteenth century generally grouped literary, political, and scientific material together, expecting their readers to understand literature and science as equally within their ambit.[13] The

[12] See for instance James A. Secord, *Controversy in Victorian Geology: The Cambrian-Silurian Dispute* (Princeton, N.J.: Princeton University Press, 1986); Secord, 'Introduction', in Charles Lyell, *Principles of Geology* (Harmondsworth: Penguin, 1997); Andrew Radford, *Thomas Hardy and the Survivals of Time* (Aldershot: Ashgate, 2003).

[13] For example, Laura Otis, 'Introduction', in Laura Otis, ed., *Literature and Science in the Nineteenth Century: An Anthology* (Oxford: Oxford University Press), p. xvii.

mathematization of scientific discourse was already well under way in some branches of science in the early nineteenth century—especially, for example, in optics and astronomy—but in many others it had not yet effectively overtaken the use of an English which, although specialized, was accessible to most general readers. Perhaps even more importantly, the separation of what the contemporary Cambridge polymath William Whewell called 'sympathies and intellectual habits' across the divide of literary culture from scientific culture was not yet entrenched, so that popularizations, summaries, and reports of front-line scientific research were a standard element of mainstream print culture.[14] The early nineteenth-century phrase for the extraordinary conjunction of enormous elite scientific progress and convulsive movements towards mass education was 'the march of mind'. The march of mind is an important trope for this book, because it stands as a reminder of the interrelationship between the period's spatial imagination and its new ways of dealing with knowledge.

'The march of mind' is a key instance of the spatial metaphors which structured access to knowledge and to the emergent public sphere in the early nineteenth century. When Keats used the phrase 'the march of intellect' in his famous letter to Reynolds in 1818, he referred (unironically) to progress towards greater knowledge: Wordsworth's greater depth relative to Milton indicates that 'there is really a grand march of intellect'.[15] But equally, the 'march of mind' could refer to horizontal progress, progress across society as well as upwards towards perfection. It expressed the contemporary idea that knowledge was spreading, was covering more territory, and reaching more people than previously. That is to say, the 'march of mind' directly invoked the questions of the ownership and management of access to knowledge that formed such an important part of early nineteenth-century cultural and political debate. Beyond this, the phrase played an important role in defining a sense of the difference between the early nineteenth century and all previous eras in British history, whether that difference were thought of as marking progress or the reverse. And it is not too fanciful to suggest that the phrase allowed the military connotations of 'marching' to be transferred into progressive civil discourse at a time when the European nations were recovering from a quarter-century of war.

14 [William Whewell], review of Mary Somerville's *On the Connexion of the Physical Sciences*, in *Quarterly Review*, 51 (March 1834), 54–68 (p. 59).
15 Keats, letter of 3 May 1818, in *The Letters of John Keats 1814–1821*, ed. Hyder Edward Rollins (Cambridge: CUP, 1958), 4 vols, I, 275–83 (p. 282).

The 'march of mind' or 'march of intellect' summed up the combination of elite intellectual progress with the beginnings of mass education. It implied the irresistibility of progress as well as its popular nature. Unlike 'the schoolmaster is abroad', Henry Brougham's similar phrase for a liberal version of the educational *Zeitgeist*, 'the march of mind' suggested what could crudely be called a 'bottom-up' approach to changes in access to education and the print media.[16] This made it peculiarly subject to condescension and downright hostility from conservative writers. One of the striking aspects of the phrase's centrality in constructing a sense of the contemporary was that it was used in ridicule by the enemies of educational reform and the spread of literacy as often as by progressives.

Of the many attacks on mass education made using the phrase 'the march of mind', Thomas Love Peacock's is now the best known. In *Crotchet Castle* (1831), the fourth of Peacock's satirical novels on contemporary ideas and intellectual life, the conservative clergyman Doctor Folliott declares himself

out of all patience with this march of mind. Here has my house been nearly burned down, by my cook taking it into her head to study hydrostatics, in a sixpenny tract, published by the Steam Intellect Society, and written by a learned friend who is for doing all the world's business as well as his own, and is equally well qualified to handle every branch of human knowledge. [...] My cook must read his rubbish in bed; and as might naturally be expected, she dropped suddenly fast asleep, overturned the candle, and set the curtains in a blaze.[17]

Peacock ridicules both parties in the march of mind: the uneducated who seek to improve their knowledge, and the educated who overestimate their own advantages and clumsily offer them to others who are unsuited for them. Near the end of the novel, Folliott focuses his attack on the impact the march of mind is having on class politics. He conflates the aspirant working class with the violent agrarian movement 'Captain Swing', and then with the Jacquerie—a rebellious peasantry. Folliott

---

[16] 29 January 1828, *Hansard*, 2nd Ser., xviii, 57–8; the phrase is often misattributed to Bentham. But it should be remembered that, in his hugely influential 'Practical Observations upon the Education of the People', Brougham emphasized the role of the populace in education reform: 'the people themselves must be the great agents in accomplishing the work of their own instruction' (*Speeches of Henry Lord Brougham...*, 4 vols (Edinburgh: Black *et al.*, 1838), IV, 99–152 (p. 103)).

[17] Thomas Love Peacock, *Crotchet Castle*, in *The Works of Thomas Love Peacock...*, ed. Henry Cole (London: Bentley, 1875), 3 vols, II, 183–285 (pp. 191–2).

insists that 'they are the same thing, under different names. [...] What was Jacquerie in the dark ages, is the march of mind in this very enlightened one.'[18] He is outraged that the spread of information to parts of society that can only translate it into envy and violence is destroying previously harmonious class relations.

Peacock was one of many who objected to the march of mind on class grounds, a line of argument which often resorted to attacking the 'relevance' of information as a means to limit access to knowledge. Many twentieth-century commentators have agreed that his attack on the educational material produced by the Society for the Diffusion of Useful Knowledge (or 'Steam Intellect Society', as he renames it) was justified by the irrelevance of such material to the lives of its readers—as in Peacock's example of the cook studying hydrostatics.[19] This appeal to utility works by assuming that different classes have different reasons for reading, and that the working classes are unlikely to have interests in scientific or other subjects outside their own immediate lives. But the evidence indicates that alongside the desire for information relevant to their trades, nineteenth-century working-class proponents of education often argued for a broad, literary, humanist education that would encourage a wider and deeper understanding of social, political, and moral issues.[20] As the radical writer Samuel Bamford put it, 'we want something more than mere intellectuality; that is already vigorous in produce, whilst SOULS lie comparatively waste'.[21] The debates around the march of mind, then, directly addressed questions of the changes in the management of knowledge precipitated by the development of a new reading public hungry for information and learning on all subjects. And they frequently did it by reference to images and tropes based on ideas of space.

---

[18] Peacock, *Crotchet Castle*, p. 278.

[19] See for instance Harold Silver, *The Concept of Popular Education: A Study of Ideas and Social Movements in the Early Nineteenth Century* (London: MacGibbon & Kee, 1965), p. 214 and more recently Alan Richardson, *Literature, Education, and Romanticism: Reading as Social Practice, 1780–1832* (Cambridge: CUP, 1994), p. 223.

[20] For a helpful introductory survey of the struggle of working-class educationalists to add political and social knowledge to the predominantly scientific and technical subjects advocated by middle-class institutions such as the Society for the Diffusion of Useful Knowledge, see Richard Johnson, ' "Really Useful Knowledge": Radical Education and Working-Class Culture, 1790–1848', in John Clarke, Chas Critcher and Richard Johnson, eds, *Working-Class Culture: Studies in History and Theory* (London: Hutchinson, 1979), pp. 75–102 (pp. 84–8).

[21] Samuel Bamford, *Passages in the Life of a Radical* (1844) (London: MacGibbon & Kee, 1967), p. 17.

The attack on the march of mind often focused on the possibilities of the phrase itself. Both sides of the argument played heavily on the spatial element of the 'march of mind' metaphor. Although Folliott's cook participates in the march of mind distinctly statically, in bed, other writers using the phrase made the most of the opportunity to play with ideas based on its conjunction of intellectual life and motion or travel. Thomas Hood, who seems to have had a jeer for every celebrity or social movement of the period, contributed to the march of mind debate an animal fable called 'The Monkey-Martyr' in which a monkey becomes a fervent activist for progress through his reading in political periodicals. The poem ends with the animal-liberationist monkey attempting to free a lion from his cage but being eaten instead. En route to this cautionary denouement, Hood describes the monkey walking down the Strand,

> striding with a step that seem'd design'd
> To represent the mighty March of Mind,
>   Instead of that slow waddle
> Of thought, to which our ancestors inclined.[22]

The 'march' metaphor allows Hood to suggest that progress is moving too rapidly to be secure in its conclusions. All the energy of the progressives is put into expediting the journey; the wisdom of the intended destination is not properly assessed. In another poem of the same period, Hood rather similarly pictures the 'March of Mind' as travelling on 'mighty stilts' only to arrive at the absurdly bathetic destination of 'a Conversazione at Hog's Norton'.[23]

It is not surprising that a comic writer like Hood exploited the opportunity of punning on the 'march' of mind. What is more surprising is that many serious writers explored the spatiality of the 'march', not in the context of pedestrianism (even on stilts), but emphatically in terms of the new technologies of travel.

In the year of the Reform Bill, a radical London speaker, Thomas MacConnell, gave an excitable lecture in which he directly connected the mechanization of travel with the irresistibility of intellectual progress:

---

[22] Thomas Hood, 'The Monkey-Martyr', in *The Complete Poetical Works of Thomas Hood*, ed. Walter Jerrold (London: Frowde and Oxford University Press, 1906), pp. 89–92 (p. 90).

[23] Thomas Hood, 'Literary and Literal', in *The Complete Poetical Works*, pp. 299–301 (p. 299).

Carriages on the rail road, and steam packets on the water, put the mind in motion as well as the body; and the public mind has received an impulse from machinery, which impulse will not *weaken* like an impulse in matter, in proportion to the extent of the sphere of its operation, but will, on the contrary, strengthen in that proportion.[24]

MacConnell's argument that mechanized transport, social enlighten-ment, and mass education were intimately connected reflected the widespread view that contemporary times marked a decisive shift from earlier, less intellectually enquiring eras. Even in the very different polit-ical climate of 1848, Mary Somerville was able to insist that 'the history of former ages exhibits nothing to be compared with the mental activity of the present. Steam, which annihilates time and space, fills mankind with schemes for advantage or defence; […] it is instrumental in bring-ing nations together, and uniting them in mutual bonds of friendship.'[25] But for less optimistic writers the link between mechanized transport and intellectual progress promised no such golden age.

Towards the end of the period on which this book focuses, Tennyson published 'Locksley Hall', a poem which echoes the languorous island descriptions of 'Enoch Arden', but expresses itself in more awkward metre and intensely colonialist sentiments. The unhappy speaker of this hugely popular poem imagines himself rejecting modern industrial soci-ety for 'some retreat | Deep in yonder shining Orient'.[26] In comparison with modern European life, the 'Orient' is fantasized as offering 'scope and breathing space':

> There methinks would be enjoyment more than in this march of mind,
> In the steamship, in the railway, in the thoughts that shake mankind.[27]

The poem strongly associates intellectual progress in Europe with the forward motion of modern methods of travel; the speaker prefers to dwell instead on the ambiguously desirable stillness and backwardness of the oriental islands. But the yearning for a retreat from progress is short-lived; by the end of the poem the speaker throws his lot in with technology and all it stands for, in a famously mistaken reference to railway tracks:

[24] Thomas MacConnell, *A Lecture on the Signs of the Times: Delivered in the Great Lecture Room of Robert Owen's Institution, Gray's Inn Road* (London: Eamonson *et al.*, 1832), p. 5.

[25] Mary Somerville, *Physical Geography*, 2 vols (London: Murray, 1848), II, p. 274.

[26] 'Locksley Hall', in *The Poems of Tennyson*, ed. Christopher Ricks (Harlow: Long-man, 1969; 2nd edn 1987), 3 vols, II, pp. 118–30 (p. 128).

[27] Ibid., pp. 123, 129, 128.

Not in vain the distance beckons. Forward, forward let us range,
Let the great world spin for ever down the ringing grooves of change.[28]

Earlier in the poem, the railway was linked to the 'thoughts that shake mankind'. Here the 'great world' now metaphorically becomes a railway train, displacing the vision of the static 'summer isles of Eden' in favour of intellectual and physical progress. [29] But despite the hectic movement there is a hollowness at the heart of this vision. The speaker joins with those conservative writers who saw the march of mind as consisting of movement for movement's sake, lacking direction and a destination. Spinning 'for ever' down, the momentum of European technological society is unstoppable but also aimless and uncontrollable.

The association of the 'march of mind' or 'march of intellect' with modern technologies of travel became so common that it was available as a subject of parody. The playwright J. B. Planché, for example, gently satirized it in a delightful one-act 'fairy extravaganza' of 1846 titled *The Bee and the Orange Tree*. The wicked ogre Ravagio is described in the list of dramatis personae as 'a violent opponent of the March of Intellect', and as the imprisoned heroine, Princess Amy, explains, he is thus equally opposed to the arrival of the railways:

> There's no escaping from these horrid brutes—
> Ravagio has a pair of seven-leagued boots,
> He'd soon look over both our naked noses;
> That's why the railroad here he so opposes;
> For, if on hear-say I may place reliance,
> The shoes of swiftness now are worn by science;
> And once the march of intellect beginning,
> No seven-league boots will have a chance of winning.[30]

Ravagio and the forces of tradition travel on foot (even if their pedestrianism is magically enhanced by seven-league boots) and so are unable to compete with the superior rapidity of technology, which symbolizes the march of intellect. Even in this playful and light-hearted

---

[28] Ibid., p. 130. Michael Robbins comments that this couplet was inspired by the opening of the Liverpool and Manchester Railway in 1830, an event which made a 'profound but erroneous impression' on Tennyson, who 'thought the wheels ran in grooves on the rails. [...] He was wrong about the 'grooves'; he was really meaning railways, not tram-lines.' (*The Railway Age* (London: Routledge & Kegan Paul, 1962; 2nd edn, Manchester: Manchester University Press, 1998), p. 14.)

[29] Tennyson, 'Locksley Hall', p. 128.

[30] J. R. Planché, *The Bee and the Orange Tree; Or, The Four Wishes: An Original Fairy Extravaganza in One Act* (London: no pub., 1846), pp. 24–5.

text, the disappearance of traditional means of travel is used to highlight the rapid, unstoppable and somehow *mechanical* advance of scientific and intellectual progress.

Throughout the early nineteenth century, the march of mind seemed to be bound up with new spatial practices, particularly, as we have seen, railway travel. But at the same time, it seemed—according to its enemies—to be uncertain of its ultimate destination. Its friends, of course, saw the situation differently: proponents of mass and adult education often argued that the spread of information through society was a way of ameliorating inequalities within a variety of key social structures, up to and including nation states, so that eventually the march of mind would put an end to the march of armies. As one influential enthusiast, Thomas Dick, put it in 1833,

That the period when a general diffusion of knowledge shall take place is hastening on, appears from the rapid progress which has been made in almost every department of science during the last half century; from the numerous publications on all subjects daily issuing from the press; [...] from the eager desire now excited, even among the lower orders of society, of becoming acquainted with subjects hitherto known and cultivated only by persons of the learned professions; and, above all, from the spirit of civil and religious liberty now bursting forth [...]. These circumstances [...] may be considered as so many preludes of a new and happier world, when intellectual light shall be diffused among all ranks, and in every region of the globe, when Peace shall extend her empire over the world,—when men of all nations, at present separated from each other by the effects of ignorance, and of political jealousies, shall be united by the bonds of love, of reason, and intelligence, and conduct themselves as rational and immortal beings.[31]

But for others, the march of mind seemed rather to be heightening inequalities, or to be heightening *awareness* of inequalities, and so to be leading to conflict in all social structures. Hazlitt was one of the writers who worried over his loyalties in the debate about the effect of the march of mind on the emotional stability and well-being of British families:

The greatest misfortune that can happen among relations is a different way of bringing up, so as to set one another's opinions and characters in an entirely

[31] Thomas Dick, *On the Improvement of Society by the Diffusion of Knowledge: or, An Illustration of the Advantages which would Result from a More General Dissemination of Rational and Scientific Information Among All Ranks* (Edinburgh: Waugh and Innes; Dublin: Curry; London: Whittaker, 1833), pp. 15–16.

new point of view. This often lets in an unwelcome day-light on the subject, and breeds schisms, coldness, and incurable heart-burnings in families. I have sometimes thought whether the progress of society and march of knowledge does not do more harm in this respect, by loosening the ties of domestic attachment, and preventing those who are most interested in, and anxious to think well of one another, from feeling a cordial sympathy and approbation of each other's sentiments, manners, views, &c. than it does good by any real advantage to the community at large.[32]

Rather like Peacock's argument that the march of mind was undermining the stability of the class system, Hazlitt's far more nuanced anxiety about the effect of intellectual progress on the family was based on a fear that traditional structures were being destroyed and not replaced: 'Happy, much happier, are those tribes and people who are confined to the same *caste* and way of life from sire to son'.[33] His point is that enlightenment is not welcome when it reveals differences among those whose happiness depends on the illusion of similarity and commonality. The forward momentum of the march of mind was contributing to the atomization of social groups and the dislocation of crucial ties of sympathy.

If the march of mind was taken by all sides to be profoundly characteristic of the age, it was a recognition of the extraordinary combination of profound changes in class politics, cultural affiliations, and educational opportunities that were taking place in early nineteenth-century Britain. Because of the ways in which these wholly new social and cultural conditions affected and were reflected in the relationship of literature and the physical sciences, the decades of the march of mind are the period on which my book focuses. By around 1850, the factors shaping the interactions between literature and the physical sciences were decisively shifting.

For one thing, relations between science and religion were about to become polarized by the publication in 1859 of *Origin of Species*. During the early part of the century, though controversy was not uncommon, there existed on the whole a comparatively harmonious relationship between mainstream English Christianity and the physical sciences. The clergyman Edward Craig's comment in 1836 that 'the laws of light, of gravitation, of electricity, and magnetism—all the disclosed arcana of nature, and all the visible splendours associated with them, lead

---

[32] William Hazlitt, 'On the Knowledge of Character', in *The Selected Writings of William Hazlitt*, ed. Duncan Wu (London: Pickering & Chatto, 1998), 9 vols, VI, pp. 271–83 (p. 279).

[33] Ibid., p. 279.

us up towards God' was typical of the generally tolerant relationship between science and religion in these decades.[34] In this atmosphere the relationship of literary with scientific culture was able to develop in ways that were impossible during the more charged decades of the later nineteenth century.

As well as the deterioration of the relationship between Christianity and the physical sciences, another factor that decisively changed the ways in which literature interacted with physics and chemistry in the nineteenth century was the marked shift away from verbal and towards mathematical language. During the first few decades of the century, chemistry adopted a symbolic language—in fact, problematically, several—but where it had applications for industry, agriculture, and so on, it still relied on verbal language. In physics, words became less and less dominant as the means of expression. By mid-century, Michael Faraday was one of the few physicists whose published work made minimal use of mathematics. Scientific papers more typically relied heavily on mathematical expression. A case in point is the development of field theory from Faraday to James Clerk Maxwell. Through the 1850s Faraday wrote a series of seminal papers outlining field theory, arguing that gravity, electricity, and magnetism are transmitted by lines of force which permeate all space. These papers were published with diagrams but with no equations; Faraday made his arguments in words. But by 1873, when Maxwell published his *Treatise on Electricity and Magnetism*, the language of field theory had shifted irrevocably away from words. Maxwell presented the results of his work in a string of immensely long and complex equations, later brilliantly simplified. Of course, since Faraday was an auto-didact with very little mathematical knowledge while Maxwell had been highly trained in mathematics at Edinburgh University, clearly part of the reason in this particular instance for their different choices of means of expression was biographical. But by the third quarter of the century it was Faraday's use of words rather than Maxwell's use of equations that was anomalous. And whatever the reasons for the shift from verbal to mathematical expression, one vital result was that the educated general reader was no longer able to follow front-line research in this aspect of physics.

Through the first half of the nineteenth century the physical sciences were professionalizing, putting themselves on a better footing

---

[34] Edward Craig, *A Lecture on the Formation of a Habit of Scientific Enquiry, delivered at the Staines Literary Institution* (Staines: Smith, 1836), p. 32.

in university curricula, developing specialisms each with its own practitioners, dialect, and organization, and establishing powerful institutional frameworks for the validation of scientific research and careers. By mid-century, the culture of science had changed profoundly, and with it the relationships of science to other kinds of culture. Professionalization and specialization meant—to a large extent—the exclusion of the general reader from scientific research, and hence a change in the ways in which words and ideas could pass between scientific and literary discourse. The years of the march of mind were the *last* time in which British literary culture had unmediated access to original work in the physical sciences. They were also the *first* time in which large-scale concerted attempts were made to give a mass readership access to both science and literature. They were an extraordinary period in British cultural history, one in which public discourse about science and literature and their relationship with each other was conducted among a uniquely broad range of social and intellectual contexts.

This book closes at the point in the cultural history of the physical sciences at which the general reader was forced to bow out of following original research. My argument is deeply influenced by the historical narrative which stresses growing scientific professionalization, specialization and fragmentation in the period.[35] While acknowledging the strength of this narrative, though, it is important to remember that part of the cultural achievement of science in this period, resulting in large part from its growing confidence and intellectual optimism, was its attempt to become all-embracing, comprehensive, and universal. Many works which had the greatest impact on the relationship of literature and science in the first half of the nineteenth century were those which gave themselves the broadest remit to investigate the nature of the physical world in a way which transcended the boundaries of the scientific disciplines. Some of these influential works went further, attempting to comprehend cultural history. Of these, the one that most impressed contemporary readers was Alexander von Humboldt's vast and magisterial work, *Kosmos*. Kosmos was perhaps the last and greatest work of this kind of magnificently ambitious, synthesizing, holistic conception of science aimed at the general reader as well as the expert, and for this reason it is a kind of end-point for this book.

---

[35] This is an argument supported by almost all historians of nineteenth-century science. See for instance David Knight, *The Age of Science: The Scientific World-view in the Nineteenth Century* (Oxford: Basil Blackwell, 1986), p. 4.

Published in Germany in 1845 though based on lectures given nearly twenty years earlier, the multi-volume *Kosmos* appeared in English translation during the late 1840s and 1850s. The book was the crowning achievement of Humboldt's lifetime of acclaimed scientific work, much of it undertaken in unexplored or little-known parts of South America and Central Asia. Trained under the great geologist Abraham Werner in the Mining Academy at Freiberg, Humboldt conducted scientific work in geology, botany, chemistry, and geography. His career is a reminder that the early nineteenth century was a producer of polymaths and of a polymathic conception of science.

Characteristically of the *Naturphilosophie*-inspired scientists of the early nineteenth century, Humboldt saw nature as a unity in which each part was fundamentally connected to all the others. In the preface to the first volume of *Kosmos* (or *Cosmos*, in English), Humboldt recalls his desire to show the interconnections between the various branches of science that describe the earth and its systems:

> while the outward circumstances of my life, and an irresistible impulse to the acquisition of different kinds of knowledge, led me to occupy myself for many years, apparently exclusively, with separate branches of science,—descriptive botany, geology, chemistry, geographical determinations, and terrestrial magnetism, tending to render useful the extensive journeys in which I engaged,—I had still throughout a higher aim in view; I ever desired to discern physical phaenomena in their widest mutual connection, and to comprehend Nature as a whole, animated and moved by inward forces.[36]

Cosmos took an overview of the current state of knowledge in all the sciences that study the planet and non-human life on it, from astronomy to meteorology and botany to morphology. But it also devoted two volumes to the cultural history of the study of nature from earliest times. The book was, as Humboldt modestly put it, 'at once scientific and literary in its aim'.[37] Humboldt was a pioneer of what would now be called ecocriticism, and spoke powerfully to mid-century British conceptions of the proper relationship of literature and science. Humboldt made it clear that he saw science and other cultural activities as complementary rather than antagonistic:

---

[36] Alexander von Humboldt, *Cosmos: Sketch of a Physical Description of the Universe*, transl. under the superintendence of Edward Sabine, new edn, 5 vols (London: Longman and John Murray, 1856), I, n.p. All quotations are taken from this English translation of Elizabeth Sabine.

[37] Ibid., III, p. 8.

As in [...] philosophy, poetry, and the fine arts, the primary aim of every study ought to be an inward one, that of enlarging and fertilising the intellect; so the direct aim of science should ever be the discovery of laws, and of the principles of unity, order, and connection, which every where reveal themselves in the universal life of nature. But by that happy connection, whereby the useful is ever linked with the true, the exalted, and the beautiful, science thus followed for her own sake will pour forth abundant, overflowing streams, to enrich and fertilise that industrial prosperity, which is a conquest of the intelligence of man over matter.[38]

For Humboldt there was no need to fear that the study of science would depriving a people 'of the life-giving breath of imagination, the arts and the literature which embellish life'.[39] On the contrary, scientific and literary advance went hand in hand. Humboldt's view was typical of the Romantic science whose fullest expression *Cosmos* was. From Wordsworth's 1802 'Preface to the *Lyrical Ballads*', which argued that 'the remotest discoveries of the Chemist, the Botanist, or Mineralogist, will be as proper objects of the Poet's art as any upon which it can be employed' and that 'Poetry is the breath and finer spirit of all knowledge; it is the impassioned expression which is in the countenance of all Science', on through to the mid-nineteenth century, the tradition in British and much Continental Romantic writing of seeing science and literature as allies, or translations of one another, was strong.[40] Though attacks such as Keats's on Newton's demystification of the rainbow sometimes came from the literary side, scientific writers in the period were less likely to launch broadsides against the intellectual impact of literature. Later in the century heated debates about the merits of literary versus scientific culture polarized the two kinds of activity, but in the early nineteenth century such questions tended to be focused on developing a reformed educational policy rather than opening up into a full-scale contest for hegemonic authority. One explanation is that while science was still emerging as a series of self-regulating specializations it could not muster enough cultural capital and confidence to attempt to destabilize its reasonably amicable political relationship with literary and other prestigious cultural manifestations. Another is that the powerful current within late eighteenth- and early nineteenth-century science that saw nature as the product of a small number of immensely powerful

---

[38] Humboldt, *Cosmos*, I, p. 39.   [39] Ibid., I, p. 38.
[40] William Wordsworth, 'Preface to the Lyrical Ballads' (1802), in *Lyrical Ballads*, ed. R. L. Brett and A. R. Jones (London: Routledge, 1963, 2nd edn 1991), pp. 241–72 (p. 260, p. 259).

and simple laws tended to promote an explanatory model that favoured unity over disjuncture and commonality over fragmentation. *Cosmos* was the last great expression of this kind of unifying Romantic science in Britain, but one in which the strength of the cultural capital of science could no longer be doubted.

Some reviews of Humboldt's extraordinary contribution were downright adulatory. *Fraser's* response to the second volume, in which Humboldt traces the cultural history of science, emphasized the moral and aesthetic qualities of the book:

We rise from the perusal of it with feelings of unmixed delight,—our views enlarged, our fancy embellished, our taste schooled, our reason fortified; [...] such a series of generous and magnificent views exhibited in such lucid order,—so much splendour of language,—so much deep research,—feelings so benevolent, and philosophy so exalted, can hardly fail to make the reader better as well as wiser, and more humane as well as more enlightened.[41]

The *Edinburgh Review*'s judgement on the first volume agreed that Humboldt had produced a work of moral as well as scientific value, but understood that value within a political context. Written by John Herschel, one of the most respected scientists in Europe and a man of strongly liberal political opinions, the review praised the unifying science of *Cosmos* as contributing to the welfare and improvement of all humanity. Through such ambitious and holistic science,

a larger and more entire conception of nature in itself may by degrees arise, and come to be recognised as the common property of humanity, the permanent and ennobling inheritance of generation after generation to the end of time.[42]

Herschel's own view had long been that science needed to be understood as having much broader concerns than strictly physical ones, and that scientific principles and methods could be applied to improving human group and individual behaviour so as to contribute 'towards the solution of the grand problem—how the advantages of government are to be secured with the least possible inconvenience to the governed'.[43] He was thus very much in sympathy with Humboldt's project for broadening the scope of 'science', both in terms of its content and in terms of its audience. Herschel wrote:

[41] *Fraser's Magazine*, 37, no. 218 (February 1848), 208–22 (p. 208).
[42] [J. F. W. Herschel], review of *Cosmos* in *Edinburgh Review*, 87 (January 1848), 170–229 (p. 179).
[43] John F. W. Herschel, *A Preliminary Discourse on the Study of Natural Philosophy* (London: Longman, 1830), p. 73.

The frame of Nature is not bounded by that narrow limit which is commonly understood by the term Physics. Life, thought, and moral and social relation, are all equally *natural*—equally elements of the great scheme of the Kosmos with matter and magnetism. The only imaginable reason why the sciences growing out of these ideas are not regarded and handled [...] as branches of natural science and inductive inquiry, is the great difficulty of arriving at true statements of facts [...]. These obstacles can only be removed by the general enlightenment of mankind [...]. [44]

But despite Herschel's optimism, the heyday of the all-encompassing, unifying Romanticism that had given science a good deal of its prestige in the cultural conversations of the past fifty years was over. Replacing it was a claim to authority that lay predominantly in the professional, specialized, and disciplinary structures of science, which could not and did not seek the same comradely relationship with literature or the same respectful mutuality with mainstream Christianity as early-nineteenth-century science had often enjoyed. *Cosmos*'s breadth of ambition marked both a beginning and an end in nineteenth-century science in Britain; it also marks the end of this book.

This book is divided into two parts. The first deals with spatial metaphors, tropes, and models used in early nineteenth-century writing about new knowledges and new conditions of knowledge. The second addresses writing about space itself, particularly focusing on three areas: pure space in geometry, new ideas about the nature of space in physics, and epic poetry dealing with cosmic space.

Chapter 1 discusses the landscape metaphors that were very frequently used to regulate and describe access to knowledge. By imaging knowledge as a series of outdoor spaces, subject to a greater or lesser degree of human intervention, and accessible to a greater or lesser range of human characters, these metaphors of gardens, wildernesses, pastoral, and picturesque landscapes staged debates about who should be able to learn what in early nineteenth-century Britain as a tussle between nature and culture. This chapter asks how spatial rhetorics were mobilized to manage access to knowledge in the face of strong pressures at once to enlarge and to control popular engagement in education. Using evidence from both popular and elite texts, the chapter explores the political implications of the particular kinds of spatial metaphor most

---

[44] [J. F. W. Herschel], review of *Cosmos*, p. 182.

widely deployed, focusing especially on arguments and anxieties about individualism, conquest, and exploration.

Landscape metaphors tended to figure knowledge as an immense tract of country through which hardy and determined learners must struggle. But many efforts were made in the early nineteenth century to picture knowledge as a harmonious, balanced system, which could be mastered more gently, often simply by detached observation, and in these attempts too spatial models were widely and variously deployed. Chapter 2 discusses some contemporary attempts to construct a model for laying out knowledge in comprehensible patterns. Opening and closing with a reminder via *Middlemarch* of some of the practical and ethical issues involved in organizing knowledge, the chapter moves on to use readings of Wordsworth's *Guide through the Lakes* and Coleridge's *Biographia Literaria* to highlight different ways of imagining knowledge as a visual spatial pattern. I focus on the hub-and-ray pattern, where the observer is placed at the centre of a series of radiating lines each formed by items of knowledge, and the aerial view, which locates the observer high above an array of information which can have a more random distribution. Each model constructs a different power relationship between the observer and the knowledge spread below or around him. Radically differently from the landscape images in Chapter 1, though, all these models aspire to present knowledge in an atemporal form—to arrange it in such a way that no time is taken in assimilating it. These models respond to the challenge from mass education by trying to take the narrative out of knowledge and to render it as dehumanized pure form.

The next two chapters explore the uses of spatial metaphor in validating and authorizing new knowledge. Chapter 3 focuses on debates about the construction of boundaries between scientific disciplines in the period. Even as the physical sciences were rapidly separating into disciplines, each with its own dialect and infrastructure, voices were being raised against the process. One powerful objection to specialization was the early nineteenth-century sense that nature constituted a single uniform field, governed by a very small number of universal laws. Carving science up into artificial disciplinary areas ran counter to the way nature itself worked. A further objection to disciplinization was that specialisms meant exclusivity and yet it was vital to keep the educated general reader abreast of the new scientific knowledge and alive to the growing confidence of scientific claims to cultural authority. The chapter shows how geographical imagery was used in this argument, with rhetoric based on borders, battlefields, nations, and kingdoms widely deployed to increase the emotional pressure on both sides.

Chapter 4 develops the analysis of the rhetoric of nations in science by looking at how international competition in European science affected English scientists' rhetorical practice. During the 1820s and particularly the 1830s, the English scientific establishment was deeply troubled by its standing among its European allies and rivals. Scientists from all parts of the political spectrum tried to describe English science in terms of English history and culture, some to argue that a profound, ancien régime-style corruption and lassitude had set in and it was time for a revolution in the old scientific order. One widely voiced concern was about language: since Latin had more or less disappeared from use as a practical lingua franca, characteristic English ineptitude with modern languages was seriously holding back English science—or so Charles Babbage and others argued. But Latin had not vanished: it had simply gone under the surface of scientific language. It still played a powerful role in maintaining science's claim to be part of a more general culture, and it could be mobilized to bolster scientific claims to authority, prestige, and heritage. This chapter focuses on a 'baptism' in early nineteenth-century British science, that of Michael Faraday's names for the process of electrolysis. This baptism was a highly conscious and deliberate attempt to deploy rhetoric which would tie new scientific ideas into a much older matrix of cultural references and associations. Literature and science studies needs to pay attention not only to the challenge from the new but to the desirability for many purposes of the old.

The second part of the book reverses the perspective, focusing on writing directly about space. Chapter 5 argues that geometry carried profound cultural significance for the early nineteenth century by representing space in its purest form. Geometry offered access to space set free from the clumsiness and limitation of human perception. Geometry's space of pure thought gave the most certain and secure rational knowledge possible and was used as a yardstick for other claims to truth and universality. Euclidean geometry had traditionally enjoyed enormously high prestige as the foundation not only for a scientific education but for any kind of systematic thought. Its authority was appropriated by many other discourses claiming cultural power. But by the early nineteenth century cracks were appearing in its status as groups with no geometrical training, and no belief in its relevance to their educational needs, denounced its prestige as a repressive outdated shibboleth. Even much more moderate writers, with vested interests in retaining geometry and other kinds of mathematical education at the heart of the British elite educational system, offered suggestions for reforming its teaching, and for making it more usefully applicable to the needs of industrial and technological workers. Geometry was the unlikely locus of important struggles in the class politics of education in the early nineteenth century: at stake was the privileging of ideal over actual space, and of abstract over practical science. Readings of contemporary writing on geometry are interwoven in this chapter with a study of Victorian geometry textbooks for both elite and artisan readers.

Chapter 6 argues that physics in the early nineteenth century was radically changing the nature of conceptions of space, essentially by dematerializing it. Contrary to the accepted account of Victorian culture as increasingly rooted in the material, the story of Victorian physics is one of jettisoning material explanations for phenomena. The contemporary success of the theory that light is a wave rather than a particle is one indication of the rise of dematerializing explanations in the period; another is the development of field theory. Field theory, particularly in its early form as propounded by Michael Faraday, collapses the distinction between matter and space, describing them instead as greater and lesser intensities of lines of electromagnetic and other forces. Field theory thus dramatically undermined all previous conceptions of space as passive container for material objects and paved the way for the energy physics of the late nineteenth century. It is perhaps a measure of the distance that studies of Victorian literature and science have yet to go that so little detailed work has been done on the ways in which ideas related to field theory were spread through literary and other culture. Evolutionary theory has received vastly more attention, but field theory, with its account of phenomena as the result of interpenetrating intensities of force, offers a hardly less exciting lens through which to view mid- and late-century literature.

The final chapter takes up the themes of spatial organization and systematization from Part I of this book and combines them with chapter 6's emphasis on dematerializing space. It reads a cluster of texts published in the 1820s and 1830s that explicitly tried to account for the spaces of the cosmos via a mixture of scientific and religious arguments, couched in verse of wildly varying quality. The huge craze for Biblical epic poems in this period has gone largely undiscussed by nineteenth-centuryists, but the epics give an extraordinary view into the relations of literature and science in the period. The chapter focuses on Creation epics, and in particular on their renderings of Chaos, the space existing before the Creation of the earth—and, according to some of the poems, still existing beyond the borders of the universe. Some poets followed Milton in representing Chaos as an unregulated, unharmonious mixture of qualities; others updated the science of Chaos by describing it as radical emptiness, the kind of pure void which *Paradise Lost*-era physics was not sure could exist. Comparisons between these verse epics and appearances of chaos in novels by Disraeli, Bulwer Lytton, and Maturin suggest that because Chaos was far less theologically important than Heaven or Hell, it served as a space of experiment for writers wanting to advance a variety of scientific and religio-scientific pet theories. The central anxiety was whether Chaos was the 'normal' state of cosmic space, and the divinely created universe was an anomaly within a great chaotic sea, or vice versa. The power balance between the two kinds of space represented the struggle between order and disorder, meaning and non-meaning. Biblical epic was thus a surprising but important site in which early-nineteenth-century writers worked out the implications of the immense task of organizing, regulating, and controlling the flood of new information being demanded by the new reading public.

PART I

THINKING WITH SPACES

1

Culture as Nature:
Landscape Metaphors and Access
to the World of Learning

A great deal of critical attention has been paid to the material factors affecting perceptions of landscape in the early nineteenth century. In particular, travel and tourism have been the subject of enormous critical interest in postcolonial Romantic and Victorian studies, extending critical investigations of landscape politics to global geography.[1] Equally, the domestic process of land enclosure has recently begun to interest literary scholars and critics—deservedly so, since the social and aesthetic implications of enclosure resonate in landscape rhetoric late into Victorian times, sometimes in the form of direct appeals on behalf of the values of the new system of land ownership, and sometimes as part of a nostalgic, elegiac art celebrating an increasingly sentimental and unhistorical picture of country life before the enclosure acts.[2] At the same time as it changed the politics of landscape, enclosure also affected cultural forms that drew on landscape and patterns of owning

---

[1] The critical literature on nineteenth-century tourism and travel is too large to be adequately represented here. Any list of major works on overseas travel would be likely to include James Buzard's *The Beaten Track: European Tourism, Literature, and the Ways to Culture, 1800–1918* (Oxford: Clarendon Press, 1993), Chloe Chard's and Helen Langsdon's collection *Transports: Travel, Pleasure, and Imaginative Geography, 1600–1830* (New Haven, Ct.: Yale University Press, 1996), and Barbara Stafford's monumental *Voyage into Substance: Art, Science, Nature, and the Illustrated Travel Account, 1760–1840* (Cambridge, Ma.: MIT Press, 1984).

[2] Enclosure was nothing new in the British landscape; it had been underway since the 1100s, but the late eighteenth and early nineteenth centuries saw two major waves of enclosures aimed at increasing the productivity of the land and efficiency of agricultural practice. G. E. Mingay's *Enclosure and the Small Farmer in the Age of the Industrial Revolution* (London: Macmillan, 1968) and J. M. Neeson, *Commoners: Common Right, Enclosure and Social Change in England, 1700–1820* (Cambridge: CUP, 1993) are useful modern histories of eighteenth- and nineteenth-century enclosure.

and managing land. Rachel Crawford's study of enclosure and poetic genre in the eighteenth century gives an illuminating insight into some of the effects of enclosure on contemporary literature.[3]

These studies do essential work on the mutual effects of material geography and nineteenth-century literary culture. But the connection between the real physical landscape and the kinds of writing I discuss in this chapter is much more tenuous, and already strongly mediated through other cultural manifestations. Although the landscape writing I examine here refers to geographical features and even to rural activities, it is nonetheless much more a product of literary and artistic traditions rather than of observation of the countryside. This kind of landscape writing is highly rhetorical, consciously artificial, a way of writing heroic or exemplary narratives or of marking out intellectual territories. And it is ubiquitous in early nineteenth-century writing about learning.

Landscape was the most fundamental and frequently occurring spatial trope in early nineteenth-century writing about knowledge. Landscape metaphors describe a branch of learning as a place, and usually locate a human character there to explore or traverse the place. In English literature the origins of this metaphor lie in the medieval dream or vision poems such as the *Romaunt of the Rose* and the opening of *Piers Plowman* in which an allegorized landscape stands for aspects of an emotional, theological, or moral idea. In the early nineteenth century these landscape metaphors were not only a resource but a kind of reflex in non-fiction prose, very widely available for deployment in all kinds of registers and discourses. They were, as Simon Schaffer puts it, 'unoriginal', belonging to no one writer in particular but mobilized in extraordinarily widespread and various cultural contexts.[4]

Landscape metaphors tended to be used at moments when writers wanted to use a shift of rhetorical gear to indicate a shift of perspective, often to a broad overview of a particular argument. In writing about factual knowledge—which was, after all, an immensely popular and varied kind of writing in the early nineteenth century—such overviews usually encompassed not only the present state of information in a particular field but also gave at least a nod to the history of the field

---

[3] Rachel Crawford, *Poetry, Enclosure, and the Vernacular Landscape, 1700–1830* (Cambridge: CUP, 2002).

[4] Simon Schaffer, 'The History and Geography of the Intellectual World: Whewell's Politics of Language', in Menachem Fisch and Simon Schaffer, eds., *William Whewell: A Composite Portrait* (Oxford: Clarendon Press, 1991), pp. 201–31 (p. 201). Schaffer describes them as unoriginal in the context of 'the scientific culture of the 1830s': p. 202.

and frequently alluded to its likely future as well. The metaphor of a landscape allowed writers to invoke a number of key tropes which structure the political relations of writer, reader, and field of knowledge and set up various kinds of emotional appeals based on response to territory which are then transferred to shape responses to intellectual endeavour.

THE INDIVIDUAL IN THE LANDSCAPE OF SCIENCE

The figures in these metaphorical landscapes in early nineteenth-century writing on knowledge tend to be solitary; it is comparatively unusual to find more than one in any given landscape metaphor. An exception to this tendency is the textbook, where learner and teacher are often both represented in the same landscape, though sometimes in shadowy form. (This situation is complicated, necessarily, in books intended for self-instruction, where the learner and the teacher are the same person, but can be addressed separately in their different roles.) William Ritchie, Professor of Natural History and Astronomy at London University, introduced his 1833 textbook on geometry, which was intended to be of '*intermediate* rank between works purely scientific and those of a popular or merely practical nature', with an elaborate landscape metaphor launched, like many others, from the classical phrase 'the royal road'.[5]

That there is no 'royal road' to geometry, though an old observation, is nearly as true as when it was first uttered, for though several new plans have been proposed, and well executed, the line of road sketched out and finished nearly two thousand years ago, is generally preferred. To shorten the road, to render it as smooth as the materials will admit, and to open to the view of the young traveller frequent glimpses of the rich domains of *science*, into which he has now entered, are objects which the author has kept steadily in view.

The pupil will also find resting-places at the most commanding points of view, with a few of the simpler kinds of gymnastic apparatus, where he may amuse himself for a little, and ascertain whether he is gaining any geometrical strength as he proceeds. To some inexperienced guides this may appear a waste of that valuable time which might be more profitably employed in advancing

---

[5] 'There is no royal road to geometry' is usually attributed to Euclid, sometimes to Menaechmus.

towards the Temple of Science. But to those who have frequently conducted youth along this path, those resting-places for amusement and exercise will appear in a very different light [...].[6]

Ritchie wants his reader to focus on the journey rather than the landscape. The only landscape features mentioned are the road, the resting-places, and the domains and Temple of Science. But his landscape is comparatively densely populated. The road is a collective work, sketched out by Euclid two millennia ago and maintained by modern educators. The traveller on the road enjoys not only the benefit of the work of these experts but also the companionship of his 'guide' (though that guide may be himself) and though we do not see him achieve the Temple of Science, nonetheless with all this help it is unlikely that he will get altogether lost or diverted. This metaphor is designed to give the reader confidence that the path he is embarking on towards geometrical adeptness is safe, reliable, and well-trodden.

Writers with a less privileged and confident attitude to learning could use landscape metaphors to emphasize instead the solitary nature of the same journey:

There are, in our day, many guides, on the road to the temple of science; but not one single conveyance, to save the exertion of a man's own powers. [...] The road is, indeed, smoothed, and swept, and levelled, and watered; and yet every inch must be travelled over, and not one pace can we advance without industry and attention.[7]

The author of this passage, a surgeon named Alfred Smith, was lecturing to the Ripon Mechanics' Institute in 1831, and the class interests of his audience are clear throughout his lecture. Smith's argument is that although in the past access to scientific information was kept from the multitude, including notably by the Roman Catholic Church, it is nonetheless everyone's right, and the nineteenth century has seen what Smith calls, with a clear allusion to his Protestant politics, a 'glorious revolution in science', making knowledge available to all.[8] In this context, his insistence on the necessarily individual and arduous

---

[6] William Ritchie, *Principles of Geometry, Familiarly Illustrated, and Applied to a Variety of Useful Purposes, Designed for the Instruction of Young Persons* (London: Taylor, 1833), n. p., Preface.

[7] Alfred Smith, *An Introductory Lecture on the Past and Present State of Science, In this Country, As Regards the Working Classes* (Ripon: Langdale; London: Longman *et al.*, 1831), pp. 24–5.

[8] Ibid., p. 23.

nature of the progress towards scientific education can be read as both an analogue of the Protestant emphasis on an unmediated and individual relationship with God via the Bible, and as a way of demystifying the learning of science by extending the desirable qualities in the self-educator to include some of the traits ('industry and attention') associated with the respectable working man. This same combination of Biblical discourse and class politics culminates in a landscape metaphor which pictures science as a kind of Eden from which the 'humbler and more numerous classes of society' have hitherto been excluded: 'high walls and impenetrable hedges were built and reared round the garden of Science, so that the multitude should never enter in to gather flowers or fruit'.[9]

Access, the central subject of Smith's lecture, is one of the key topics about which nineteenth-century writers asked questions by using landscape metaphors. Who is allowed into the landscape of knowledge? What are its borders? Where is the best place from which to survey it? As Smith's argument indicates, the contemporary debate about class and education was necessarily concerned with the role of the individual in the collective work of advancing human knowledge. Two years after Smith's lecture, in an address to a very different but still socially and intellectually mixed audience—the 1833 meeting of the British Association for the Advancement of Science—the Cambridge professor William Whewell addressed this question of the individual's part in science. Like Smith and many other writers on the politics of knowledge in this period, Whewell used an extended landscape metaphor, beginning—like Ritchie's—with the royal road trope:

There is, as was long ago said, no royal road to knowledge—no possibility of shortening the way, because he who wishes to travel along it is the most powerful *one*; and just as little is there any mode of making it shorter, because they who press forward are *many*. We must all start from our actual position, and we cannot accelerate our advance by any method of giving to each man his mile of the march. Yet something we may do: we may take care that those who come ready and willing for the road, shall start from the proper point and in the proper direction,—shall not scramble over broken ground, when there is a causeway parallel to their path, nor set off confidently from an advanced point when the first steps of the road are still doubtful;—shall not waste their power in struggling forwards where movement is not progress, and shall have pointed out to them all glimmerings of light, through the dense and deep screen which divides us from the next bright region of philosophical truth.[10]

Whewell's metaphor can be read as a rebuke to Alfred Smith and his ilk among the educationally unprivileged classes who were urgently demanding access to scientific knowledge. Whewell, eminent among the 'gentlemen of science' who ran the British Association in its early days, sees the struggle for learning very much from the perspective of the guide rather than the learner. 'We' perceive the landscape a great deal better than does the neophyte seeker after science, and neither individual privilege nor mass pressure will equalize the balance of power between us and them. But Whewell and Alfred Smith agree in emphasizing the solitary nature of the work to be done by the learner: a scientific education cannot be achieved except by the efforts of the individual. Toil, exertion, and dedication are required, and the only reward Whewell predicts is the 'glimmerings' of unseen truths.

Other contemporary writers used landscape metaphors in a much more optimistic way to describe the exhilaration and potentially overwhelming possibilities of entering the educated classes. John Mann was a member of the Royal College of Surgeons and of a private self-improvement circle. In the same year as William Ritchie published his geometry textbook, Mann published a lecture he gave to this circle, titled *A Glance at the Objects of Thought, or, A Concise and Systematic View of the Different Branches of Human Knowledge*. Mann's pamphlet was a very minor example of a much larger class; it was one of a great many works of widely varying degrees of ambition published in the early nineteenth century attempting to systematize, methodize, or otherwize make sense of the burgeoning specializations within human knowledge, seeking to counter the decentralizing effect of the development of disciplines. Mann uses a landscape metaphor to describe not only the current state of human learning but especially to register his mixed emotional response to the scale of the task ahead:

We will, then, if you please, attempt to take up the mind to some lofty eminence, from which it may cast a bird's-eye glance upon the field of human knowledge; and if it has hitherto been accustomed to be confined in some narrow corner of the world, where its views have become as limited as its habitation, and its faculties rendered torpid simply by the want of employment, then we would

[10] William Whewell, *Address Delivered in the Senate-House at Cambridge, June XXV, MDCCCXXXIII. On the occasion of the opening of the third General Meeting of the British Association for the Advancement of Science* (Cambridge: Smith, 1833), pp. 3–4.

hope that by being raised to a more elevated position, to the inhalation of a more free and open atmosphere, and to the contemplation of an expanded and ever-expanding landscape, it may at the same time increase its desires and enlarge its capacity for knowledge, render its constitution more healthy and its faculties more active. But if it happen that this view we propose to take should, after all, seem imperfect and uninteresting, then we beseech you to attribute it only to the manner in which it has been presented, and not to any want of interest in the subject itself. Say not of the landscape that it is bare of beauty, but rather believe that the atmosphere was misty, or that the eyes of your guide were dim.

In entering upon so extensive a field it might well be supposed that the multiplicity of objects presented to our view would produce bewilderment, and from that bewilderment nothing could result except a confusedness of vision which would leave the mind perplexed rather than informed, and wearied with the extent of the prospect rather than stimulated to exertion by its beauty or variety. Perhaps, indeed, under this confused and misty perception of things, the whole field of human knowledge might seem like the realms of Chaos and Old Night, rather than as being the kingdom of nature's harmony; the display of heaven's wisdom, the field where are found the products of Mind, the treasures of Memory, and the flowers of Imagination.[11]

Consolingly, Mann reassures his audience that 'it will hardly be so, if in our researches we allow ourselves to be guided by Method, by the aid of which confusion may give place to order, and bewilderment be changed for distinctness and delight'.[12] Unlike Ritchie, Mann does not overplay the allegory of his landscape metaphor. Though he begins with an admission of the self-consciousness of his use of metaphor ('if you please'), and employs the standard trope of the high vantage point, Mann goes on to use the image of landscape to do more complicated work than Ritchie's deterministic, teleological journey towards scientific proficiency. Instead of confidently commanding the landscape below him, Mann's imaginary spectator is liable to misperceptions and inappropriate emotional responses. Where Ritchie is interested in tracking progress in geometry, a single area of knowledge, Mann's scope is much broader and his landscape metaphor allows him to acknowledge both the excitement and also the anxiety that must have felt by many nineteenth-century readers confronted with new access to the immensity of human learning.

[11] John Mann, *A Glance at the Objects of Thought, or, A Concise and Systematic View of the Different Branches of Human Knowledge* (London: Roake and Varty, 1833), pp. 3–4.
[12] Ibid., p. 4.

Nineteenth-century working-class autobiographies often attest to this experience. David Vincent, John Burnett, and others have done a great deal to bring to light the subjective experience of the autodidact by investigating autobiographical writing undertaken by working-class men and women later in life.[13] There is much less evidence, though, about the experience of autodidacticism, particularly scientific autodidacticism, by people undergoing the experience at the time at which they write. One very valuable source of this kind is Michael Faraday. Unlike most of the aspirational young men Mann, Ritchie and countless others were addressing, Faraday wrote a great deal about the process of self-education while he himself was undergoing it; and later in life, when he had become a central figure in nineteenth-century accounts of the value of scientific education for the working-class, he published influential essays on education more generally.

Even as a very young man, Faraday wrote at length about the emotional as well as the intellectual aspects of education and self-education. His juvenile essays are almost unknown beyond a small circle of historians of science, but they deserve to be read more widely. From 1818 to 1819 Faraday belonged to an essay-circle consisting of five young men who met monthly to read and criticize one another's literary compositions. Their hope was to learn a gentlemanly writing style which they felt would improve their employment opportunities and their confidence in social situations. The essays and poems produced by this group amount to an almost unique corpus of evidence about early nineteenth-century artisan self-help writing in the capital.

Faraday's essays, unlike most of those of his friends in the essay-circle, are frequently about the need for self-improvement and the means to achieve it. They take a quasi-psychological approach to the question, addressing topics such as 'the Pleasures and Uses of the Imagination', and 'Imagination and Judgement'. Given the ubiquity of landscape metaphors in contemporary texts discussing access to knowledge, and the fact that the members of this essay-circle aspired not to originality but to conventionality, it is not at all surprising that Faraday repeatedly uses landscape metaphors to explore moral, practical, and emotional

---

[13] The three volumes of *The Autobiography of the Working Class: An Annotated, Critical Bibliography*, ed. John Burnett, David Vincent and David Mayall (London: Harvester, 1984–9), are an invaluable source for such autobiographies. For an influential discussion of a number of the autobiographies listed, see David Vincent's *Bread, Knowledge and Freedom: A Study of Nineteenth-Century Working-Class Autobiography* (London: Europa, 1981).

questions about self-education. Because this writing was intended to school them in a repertoire of standard literary conventions rather than to be autobiography, the metaphors Faraday adopts in these essays are pastoral and picturesque, with no trace of the cityscapes he and his friends really knew.

In 'On Mental Education', the essay Faraday wrote about education nearly forty years later, when he was an elder statesman of the European scientific world, he presented learning as a matter primarily of improving the judgement through 'mental discipline', 'teaching the mind to resist its desires and inclinations, until they are proved to be right', 'patience and labour of thought', and 'humility'.[14] But in his early pieces, he writes warmly about how the sheer scale of the task can seem overwhelming to the inexperienced learner. In one important passage he describes the world of knowledge that opens up to the self-educator as both beautiful and unmanageable:

The mind wanders about as if lost in the extent of country to which it has been so lately admitted strange to every thing around it has only the inclination to become acquainted without the power to decide where to commence its system of research It is misled with the extended view in the distance it wanders through the mazy valleys of knowledge and is thunderstruck at the towering hills of science. Here many are lost their spirits fail them they believe it is only for birds of stronger wing to attempt the boundless flight. But we must return back to the door which admitted us and a shorter view with modest but persevering exertions.[15]

Here Faraday uses landscape allegory to attempt to impose some order on the array of possibilities and demands facing the autodidact. Although the everyman character is lost, the narrator understands the landscape enough to be able to identify its constituent parts and give their relative positions. But the narrative voice is that of the experienced traveller rather than that of the landowner or the improver of the estate. Throughout his early essays, Faraday argues that he belongs in the realm of learning, but his landscape metaphors suggest that he recognizes that a combination of class, property rights, and aesthetics shapes that realm in ways that make it difficult for him to inhabit it comfortably. John

[14] Michael Faraday, 'On Mental Education', in *Experimental Researches in Chemistry and Physics* (London: Taylor and Francis, 1859), pp. 463–91 (pp. 475, 477, 484, 485; all quoted words italicized in original).
[15] 'On Imagination and Judgement', in 'Mental Exercises', Royal Institution, Faraday MSS, F.13.A, pp. 22–7 (pp. 25–6). The lack of punctuation adds drama and rapidity to the description and is typical of Faraday's writing in this period.

Mann's *A Glance at the Objects of Thought* pictures a learner 'perplexed' and 'wearied' by the expanse of the landscape of knowledge. Faraday's somewhat happier characters experience pleasure in this landscape, but neither employment nor home. His point of view here is that of the tourist: but an unprivileged tourist, who travels on foot and without a guide.

Faraday's experience of actual travel was much broader than those of most young men of his class outside the army or navy. During the last years of the Long French Wars he travelled extensively in Continental Europe as valet and assistant to Sir Humphry Davy, whose party was granted special passes by Napoleon in recognition of Davy's scientific standing. Faraday took his responsibilities as a tourist very seriously, trying to learn the languages of the countries he visited, doing his best to see the sights in each city, and keeping a detailed journal of his impressions. In his recent biography of Faraday, James Hamilton describes this journal as 'one of the fullest and most exciting travel documents of the period'.[16] But none of this unique experience of travel is reflected in the landscape metaphors he constructs in his literary essays.

Instead, Faraday's strongest spatial metaphors in these essays draw their detail from the well-used stocks-in-trade of late eighteenth-century landscape description, particularly in the picturesque mode. He resorts frequently to the typical elements of picturesque landscape painting—hills, waters, woods, ruins. In his essay 'On the Pleasures and Uses of the Imagination', for instance, he describes how a person who has cultivated his mind can enter imaginatively into the materiality of a landscape: 'he will penetrate into the recesses of a wood, bathe in the waters, explore the ruins of old castles, or disappear behind some heath clad hill'.[17] The interweaving of aesthetic modes (literature, painting, even agriculture) was characteristic of the appreciation of the picturesque, and Faraday had so thoroughly absorbed the fashionable taste which dictated that both real and imaginary landscapes be judged according to the same generic elements and qualities, that it is impossible to determine whether the scene he describes here is a real tract of country or a painting.[18]

---

[16] James Hamilton, *A Life of Discovery: Michael Faraday, Giant of the Scientific Revolution* (London: Random House, 2002), p. 51.

[17] [Michael Faraday], 'On the Pleasures and Uses of the Imagination', in 'Mental Exercises', pp. 39–47 (p. 47).

[18] Richard Payne Knight commented, for example, that 'a person conversant with the writings of Theocritus and Virgil will relish pastoral scenery more than one unacquainted

But by 1818 the elements of the picturesque landscape were very well-known: so why should Faraday's wanderer in the country of knowledge feel bewildered in the face of this familiar setting? The picturesque landscape aims to inspire emotion in the observer or traverser; and unlike the classical style, with its open views and empty spaces, picturesque trades on the provocation of surprise and disorientation, even when that surprise and disorientation are conventional. Thomas Love Peacock highlights the absurdity of the conventionality of these responses to the picturesque when he has the fashionable Mr Milestone ask whether the aesthetic quality of 'unexpectedness', advocated by Mr Gall as crucial to landscaped gardens, can be felt 'when a person walks round the grounds for the second time'.[19] But Faraday's account of bewilderment in the landscape is not lessened by his recognition of the stock objects of picturesque landscape art and literature. Instead, such a conventional landscape is the only one in which he can imagine himself feeling this conventional bewilderment.

The habit of reading real and imaginary landscapes via the same aesthetic criteria coloured Faraday's experience of domestic tourism. In 1819, the year the essay-circle folded, Faraday, aged 27, made a walking tour of north Wales with his friend Edward Magrath. In his journal of this tour he filtered descriptions of real places through standards derived from fashionable art. For example:

The scenery was at first moderately pretty being composed of high hills or rather mountains with a small stream of water falling in a long succession of cascades before us but wood was wanting and that is with difficulty replaced in a landscape by any other feature.[20]

This landscape is appreciated as a picture: the lack of timber is unfortunate for aesthetic, not economic or environmental, reasons. Similarly, Faraday records even human figures in the landscape as noteworthy primarily for their aesthetic function: 'lively little cattle with myriads of sheep now and then diversified the general monotony and a turf cutter

---

with such poetry' (*An Analytical Inquiry into the Principles of Taste*, 2nd edn (London: Payne and White, 1805), p. 150; quoted in Malcolm Andrews, *The Search for the Picturesque: Landscape Aesthetics and Tourism in Britain, 1760–1800*, (Aldershot: Scolar, 1989), p. 4).

[19] Thomas Love Peacock, *Headlong Hall*, in *The Works of Thomas Love Peacock ...*, ed. Henry Cole, 3 vols (London: Bentley, 1875) I, pp. 31–2.

[20] Dafydd Tomos, ed., *Michael Faraday in Wales: Including Faraday's Journal of his Tour through Wales in 1819* ([Denbigh (?)]: Gwasg Gee, [1972]), pp. 54–5.

or a peat digger here and there drew the eye for want of a better object'.[21] Several studies have drawn attention to the picturesque's tendency to dehumanize landscape.[22] Faraday does at least recognize and name the occupations of these Welsh labourers. But he seems to equate them with cattle and sheep in terms of status and interest, and he aestheticizes both animals and people as elements of a picture rather than elements of an economy.

Recent critics have tended to underline the elite politics of the picturesque.[23] Among the features of the literary picturesque, for instance, is a reliance on familiarity with a literary canon that privileges classical literature, a kind of education which Faraday and artisan self-improvers of his generation lacked. Malcolm Andrews has described the taste for the picturesque as a kind of hobby purchase, 'an extra, expensive piece of intellectual equipment to take into the field'.[24] Stephen Copley and Peter Garside, on the other hand, see it as something to be imposed on, rather than taken into, the field. They emphasize the role played by the picturesque in the process of colonization within, as well as outside, Europe's boundaries. Copley and Garside argue that in the Scottish highlands, 'the combination of political repression, economic exploitation, and aesthetic sentimentalisation of the Scottish landscape in the early nineteenth century clearly renders the Picturesque "invention" of the region a hegemonic cultural manifestation of the English colonising presence'.[25] A similar point could be made about Faraday's colonial relationship with the Wales he explored on his walking tour. Wales, particularly the north, was 'invented' by the English picturesque tourist, very much as the Scottish highlands were 'invented'. By the start of the

[21] Ibid., p. 55.

[22] Perhaps the most influential have been Raymond Williams, *The Country and the City* (London: Chatto & Windus, 1973) and John Barrell, *The Dark Side of the Landscape: The Rural Poor in English Painting 1730–1840* (Cambridge: CUP, 1980).

[23] The elite politics of the picturesque should not be treated as wholly dominating the discourse, however. Tim Fulford reminds us that the picturesque was capable of expressing the views of those writers who, 'vindicating the taste of gentry and nobility, nevertheless were haunted by visions of natural and social forces beyond the limits of that taste'. In particular, William Gilpin's version of the picturesque 'articulated [...] his sympathy with the rebellious power of the lower orders who resisted the authority—and the aesthetics—of gentlemen' (*Landscape, Liberty and Authority: Poetry, Criticism and Politics from Thomson to Wordsworth* (Cambridge: CUP, 1996), p. 15.

[24] Andrews, *Search for the Picturesque*, p. 3; see p. 4 for 'the elitism of the Picturesque'.

[25] Stephen Copley and Peter Garside, 'Introduction', in *The Politics of the Picturesque: Literature, Landscape and Aesthetics since 1770*, ed. Copley and Garside (Cambridge: CUP, 1994), pp. 1–12 (pp. 6–7).

nineteenth century, around twenty years into north Wales's phase of popularity among picturesque tourists, 'the general verdict that Wales seemed at least a century behind England changed from being an insult and a deterrent to being a positive incentive for tourists'.[26] Faraday's tour followed one of the most common routes through Wales, travelling north from Devil's Bridge through Dolgellau, the Vale of Festiniog and Caernarvon, and then turning east to Bangor and finally Llangollen, taking in Cader Idris, the Menai Straits, and the River Dee on the way.[27] In his route, his expressed responses and his aesthetic criteria, Faraday was very much of his time in his understanding of the spaces of domestic tourism.

The fact that a London artisan like Faraday was able to adopt the picturesque so wholly indicates how far down the social scale the mode had penetrated. Although it had originally had a close and real connection with classical literature and an elite education, by the second decade of the nineteenth century this connection had become conventional, opening picturesque appreciation to many for whom its literary roots were inaccessible, and this opening necessitated alterations in the politicization of landscape representation.

The aesthetics and political values of the picturesque had originally been associated at least in part with the preservation of the rights of a social elite: Ann Bermingham's succinct formulation captures what we might think of as the mission statement of the picturesque: 'where power was, there beauty shall reside'.[28] However, the growing popularity of the picturesque made these aesthetics and politics available to a fairly wide public, including people like Faraday, whose socioeconomic interests were opposed in almost all regards to those of the landowners.[29] It was characteristic of Faraday's efforts to conform to contemporary gentlemanly aesthetic tastes that he adopted the picturesque as his means of interpreting landscape, and characteristic of the effect of the picturesque that he applied it both to real and imaginary landscapes.

[26] Andrews, *Search for the Picturesque*, pp. 114, 150.

[27] See Andrews, *Search for the Picturesque*, pp. 114–50, for a detailed and illustrated account of the picturesque scenes of this part of Wales.

[28] Ann Bermingham, *Landscape and Ideology: The English Rustic Tradition, 1740–1860* (London: Thames and Hudson, 1986), p. 83. Bermingham points out that 'power' in this context meant the elite of both the older, quasi-feudal and the newer, industrial agricultural economies.

[29] Stephen Copley and Peter Garside note the difficulty of dealing with the variety of '"high cultural" and "popular" contexts in which Picturesque terminology is used' (Copley and Garside, *The Politics of the Picturesque*, p. 2).

The popularization of the picturesque brought the viewpoints of the landowner and the tourist closer together as both tried to perceive landscape through a given set of pre-determined aesthetic criteria. Faraday's landscape metaphors about education exemplify the way in which this loosening of the distinction between owner and spectator complemented the loosening of the boundary between the traditionally educationally privileged and those arriving at education for the first time. Imaginary landscapes were a major part of the standard rhetoric of education in the early nineteenth century, and where they adopted the tropes of the picturesque they could be used to adopt the politics of that mode to the new politics of educational access.

But of course other kinds of aesthetics and politics were also available via landscape metaphors. Faraday's later writing, and writing by others about him, tends to use a range of more masculine, active, and controlling kinds of landscape metaphor. Some of these landscape tropes seem to have had considerable effect on the public conception of the scientist which developed around the myth of Faraday; and this myth in its turn had a powerful role in the development of a Victorian public ideal of the scientist and of the self-educator.[30]

By mid-century, both scientific triumphalism and imperialist ideology were finding the territorial and geographical possibilities of landscape metaphors very useful for expressing claims of possession and supremacy. Iconic figures of science (and other fields of endeavour) were envisaged

[30] See for instance Alice Jenkins, 'Spatial Rhetoric in the Self-presentation of Nineteenth-century Scientists: Faraday and Tyndall', in Crosbie Smith and Jon Agar, eds., *Making Space for the History of Science* (Basingstoke: Macmillan, 1998), pp. 181–91. The Duke of Somerset called Faraday 'a Hero of chemistry' as early as 1835, noting that 'there is something romantic and quite affecting in such a conjunction of poverty and passion for science' (quoted in Simon Schaffer, 'The History and Geography of the Intellectual World', p. 226), and right through the Victorian period most commentators on Faraday's biography agreed with the Duke. In 1859 Samuel Smiles famously lauded Faraday as 'the son of a blacksmith, [who] now occupies the very first rank as a philosopher, excelling even his master, Sir Humphrey Davy, in the art of lucidly expounding the most difficult and abstruse points in natural science' (*Self-Help; with Illustrations of Character, Conduct, and Perseverance*, ed. Peter W. Sinnema (Oxford: Oxford World's Classics, 2002), p. 24). It is not too much to call Faraday an icon of Victorian self-education, and he was certainly an icon of Victorian science. For useful studies exploring Faraday's iconicity see Geoffrey Cantor, 'The Scientist as Hero: Public Images of Michael Faraday', in Michael Shortland and Richard Yeo, eds, *Telling Lives in Science: Essays on Scientific Biography* (Cambridge: CUP, 1996), pp. 171–94; Graeme Gooday, 'Faraday Reinvented: Moral Imagery and Institutional Icons in Victorian Electrical Engineering', *History of Technology* 15 (1993), 190–205; and Iwan Rhys Morus, *Michael Faraday and the Electrical Century* (Cambridge: Icon, 2004), esp. pp. 216–21.

in iconic landscapes, demonstrating by their dominance over these landscapes their claims to discovery, ownership and license to exploit. Charles Daubeny's compliment to Faraday's original patron, the recently deceased Sir Humphry Davy, uses just this technique :

For the attainment of that eminence, whence these prospects of nature have disclosed themselves, we are mainly indebted to the last generation of chemists, and above all to the illustrious President of the Royal Society, whose death science has lately had to deplore; but, considering the unexampled rapidity with which one discovery of his succeeded another, and the eagerness with which he continued to press forwards into new regions of inquiry, it is not wonderful, that he should have left a large portion of the field that he had traversed unexamined, its boundaries ill-defined, and its treasures in a great degree unexplored.[31]

Daubeny's praise for the heroic Davy is an intervention in the contemporary argument about the intersection of class with the competing claims of pure and applied science. As Daubeny explains, he hopes to rid his readers of the prejudice that 'the researches of the present age are less calculated to train and invigorate the understanding, than those of the ancients, because they are in general more directed towards objects of a practical, and therefore, as is supposed, of a less intellectual description'.[32] Picturing Davy as an intrepid explorer who did not stop to exploit his territorial discoveries allows Daubeny to emphasize the gentlemanliness of the leaders of chemistry. He shows that he expects the readers of his own introduction to atomic theory to have a literary education by quoting from Abraham Cowley: 'For life did never to one man allow | Time to discover worlds, and conquer too.'[33] But, Davy having forged ahead with his discoveries, it is now time to follow and undertake the less visionary work, and for this task Daubeny, his class interests in mind, shifts to a different landscape metaphor, one with pastoral rather than commercial implications:

It is for the present race of chemists therefore, to fill up the magnificent outline that has been traced out for them by their predecessors, and to cull the fruits thus brought within their reach; neither will the present Treatise be thrown away, if

---

[31] Charles Daubeny, *An Introduction to the Atomic Theory, Comprising a Sketch of the Opinions Entertained by the Most Distinguished Ancient and Modern Philosophers With Respect to the Constitution of Matter* (Oxford and London: Murray, 1831), p. x.
[32] Ibid., p. vii.
[33] Ibid., p. x. The Cowley quotation is from his 1663 poem 'To the Royal Society'.

it succeeds in attracting some fresh labourers to the harvest, by contributing to make known its abundance and extent.[34]

As these examples from Daubeny's book illustrate, two major kinds of landscape metaphor structured rhetoric surrounding the image of the nineteenth-century scientific practitioner: those based on ideas of territorial discovery and conquest, and those based on references to farming, gardening, and generally tending the land. Faraday, one of the great scientific rhetoricians as well as scientific investigators of the first half of the nineteenth century, frequently made rhetorical use of both kinds of metaphor.

When discussing science as a collective enterprise he sometimes used what we might call the 'united empire of knowledge' trope. For instance, in his essay on the conservation of force, a principle which argues the essential unity of nature, he stressed the essential unity of the landscape of science. The principle of the conservation of force 'should be the more earnestly employed and the more frequently resorted to when we are labouring either to discover new regions of science, or to map out and develop those which are known into one harmonious whole'.[35] But in his self-presentation he was careful to use humbler landscape metaphors. In the same paper on the conservation of force, Faraday compared people with 'high and piercing intellects' to

those persevering labourers (amongst whom I endeavour to class myself), who, occupied in the comparison of physical ideas with fundamental principles, and continually sustaining and aiding themselves by experiment and observation, delight to labour for the advance of natural knowledge, and strive to follow it into undiscovered regions.[36]

It was difficult to keep the different kinds of landscape metaphor—of agriculture, of empire, of exploration—separate. They often tended, in practice, to merge into one another. Here metaphors of agricultural labour drift into images of exploration, though it is characteristic of Faraday to present himself initially at least in terms of the agricultural metaphor.

The image of the labourer in the field was sanctioned and strengthened for Faraday and many other writers by its Biblical, and particularly

[34] Daubeny, *An Introduction to the Atomic Theory* (1831 edn), pp. x–xi.
[35] Michael Faraday, 'On the Conservation of Force', in *Experimental Researches in Chemistry and Physics*, pp. 443–63 (p. 459).
[36] Ibid., p. 443.

Dissenting, connotations.[37] The mysterious Scriptural parable of the hired labourers, all of whom were paid the same wage regardless of the length of their working day, is at the heart of the New Testament's exposition of the final separation of the saved from the lost on Judgement Day, and related allegories are very common in English Protestant religious writing.[38] The identification of the husbandman, agriculturalist, or horticulturalist with the Christian coloured English Protestant landscape metaphors well into the nineteenth century.

Picturesque and agricultural landscape metaphors imply quite different attitudes to access to knowledge. Where the landscapes in Faraday's early essays construct the observer as a tourist, exploring at will or by whim the attractive and dramatic scenery of systematized knowledge, these later agricultural landscapes are imbued with values associated with work. The contrast is that between consumption and labour, and between solitary and collective enterprise. The aesthetic landscapes can accommodate only one tourist at a time. The effect of landscape (both in rhetoric and actual experience) would be spoilt if carriage-loads of visitors were exploring it together, as Wordsworth pointed out in his protest against railways in the Lake District. But agricultural landscape metaphors can envisage a team of workers co-operating, many people together gaining access to new knowledge.

THE LIMITS OF THE LANDSCAPE OF SCIENCE

Thomas Sprat's mid-seventeenth century *History of the Royal Society* set out the ideological and methodological mission of this bastion of the British scientific establishment. Sprat used a landscape metaphor to explain how science's task is to bring humanity closer to God by heaping up scientific information until we gain a high enough station to be able to view God. He alludes to the tower of Babel but distinguishes triumphantly between those times and his own:

[37] Geoffrey Cantor's *Michael Faraday: Sandemanian and Scientist* (Houndmills: Macmillan, 1991) has been very influential in locating Faraday's science in the context of his Sandemanianism, with its strong emphasis on the Bible.

[38] Matthew 20.1–16. One example is John Flavel's *Husbandry Spiritualized*, published originally in the seventeenth century, but extremely popular and continually reprinted in various editions until at least the mid-Victorian period: *Husbandry Spiritualized; or, The Heavenly Use of Earthly Things [...]* (London: Boulter, 1669).

This is truly to command the world; to rank all the *varieties*, and *degrees* of things, so orderly one upon another; that standing on the top of them, we may perfectly behold all that are below, and make them all serviceable to the quiet, and peace, and plenty of Man's life. And to this happiness, there can be nothing else added: but that we make a second advantage of this *rising ground*, thereby to look the nearer into heaven: An ambition, which though it was punish'd in the *old World*, by an *universal Confusion*; when it was manag'd with *impiety*, and *insolence*: yet, when it is carried on by that *humility* and *innocence*, which can never be separated from true knowledg; when it is design'd, not to *brave* the Creator of all things, but to *admire* him the more: it must needs be the utmost perfection of *humane Nature*.[39]

For Sprat, scientific knowledge is power: it allows humans to take their place 'on the top of' the rest of creation and even to see confidently into the realm of the divine. The 'rising ground' on which science stands gives us the advantage of a breadth of vision to which Sprat sees no limits.

A hundred and fifty years later, early nineteenth-century writing on science frequently discussed the question of whether there was a limit to the knowledge which would ever be achievable by humans. It would be a misinterpretation to read many of these discussions as symptomatic of a lack of confidence in scientific practice or method. Rather, the suggestion that human reason might at some point face a barrier beyond which it could not proceed was evidence of a growing sense of identity and optimistic purpose in professionalizing science. It allowed writers proposing the authority of science to negotiate gracefully the relationship of science with other kinds of knowledge, particularly religion; and in turn when religious writers used it to try to claim back some of that authority it generally served only to acknowledge the solidity of science's claims. The image of a boundary at the edge of science was an element in the pattern of polite relations of literature and science, though one increasingly deployed by theological rather than scientific writers as the century progressed.

Scientific writers used a wide variety of landscape metaphors to explore the idea of the possible boundaries of science. In the opening chapter of his 1830 *Preliminary Discourse*, for instance, John Herschel writes about the task of the natural philosopher as rather like that of Moses, who went up from the plains of Moab to the summit of Mount

---

[39] Thomas Sprat, *The History of the Royal-Society of London, for the Improving of Natural Knowledge* (London: Martyn & Allestry, 1667), pp. 110–11.

Pisgah, and was shown the land of Gilead, but was not allowed to enter it:

all the longest life and most vigorous intellect can give [man] power to discover by his own research, or time to know by availing himself of that of others, serves only to place him on the very frontier of knowledge, and afford a distant glimpse of boundless realms beyond, where no human thought has penetrated, but which yet he is sure must be no less familiarly known to that Intelligence which he traces throughout creation than the most obvious truths which he himself daily applies to his most trifling purposes.[40]

Herschel pairs the 'boundless realms' beyond science's grasp with the 'frontier of knowledge' on which we stand. But though 'boundless' faintly raises the suggestion of a 'boundary', that is not what Herschel offers: a frontier is not after all the same thing. Similarly, though no human thought 'has penetrated' beyond this frontier, Herschel does not say that it cannot be penetrated. His metaphor is ambiguous and respects the territorial claims of both science and religion.

Herschel's friend and admirer Mary Somerville has rather more literal 'boundless realms' in mind when she writes in *On the Connexion of the Physical Sciences* that astronomy inculcates humility

by showing that there is a barrier which no energy, mental or physical, can ever enable us to pass: that however profoundly we may penetrate the depths of space, there still remain innumerable systems, compared with which these apparently so vast must dwindle into insignificance [...].[41]

Somerville subverts the religiously orthodox trope of a limit to science. Instead of suggesting that human reason is qualitatively unfitted to plumb the ultimate mysteries of the divine creation, instead she argues that the problem is one of relative scale: the universe is so large that 'not only man, but the globe he inhabits—nay, the whole system of which it forms so small a part,—might be annihilated, and its extinction be unperceived in the immensity of creation'.[42] Not only does Somerville leave the divine out of this account of the power of human science, but she does not acknowledge here the presence of God to 'perceive' the fate of humanity.

---

[40] John F.W. Herschel, *A Preliminary Discourse on the Study of Natural Philosophy* (London: Longman, 1830), pp. 6–7. See Deuteronomy 32.48–52 for the prophecy about Moses and Deuteronomy 34.1–5 for its fulfilment.

[41] Mary Somerville, *On the Connexion of the Physical Sciences* (London: John Murray, 1834), p. 4.

[42] Ibid., p. 4.

More conventionally, though, the trope of the limit of science was often deployed to draw a boundary line between the province of science and that of religion. In 1849, for instance, a Glasgow clergyman, Thomas Chalmers, published a sermon on *The Efficacy of Prayer Consistent with the Uniformity of Nature*, in which he argued that there was a limit beyond which 'we cannot trace the pathway of causation—not because the pathway ceases, but because we have lost sight of it—having now retired from view among the depths and mysteries of an unknown region, which we, with our bounded faculties, cannot enter'.[43] Beyond this limit, in 'the region of faith', Chalmers asserted the jurisdiction of religion rather than that of science.[44]

These boundary metaphors played an important part in the developing authority claims of science in the early part of the nineteenth century. As well as being involved in the negotiation of diplomatic relations between science and Christianity, they were used to carve up the political landscape of science, and more generally of education.

We might take as an example the metaphor of the 'frontier of knowledge' in Herschel's *Preliminary Discourse*, quoted above. Herschel's use of landscape metaphors was affected by his reformist politics, because they had a deep impact on his conception of what we might think of as the land economy of science: who can acquire, possess and gain access to scientific knowledge. The first edition of Herschel's *Discourse* was published at a time of great controversy over the future of some of the great institutions of British public life. The impetus to reform the more blatant abuses of the Parliamentary electoral system had a counterpart in the scientific sphere in the attempted reform of the Royal Society during the early 1830s. Herschel was profoundly involved in this attempt as the reformist faction's candidate for President of the Society in opposition to the Duke of Sussex in the election of 1830, the year in which he published the *Preliminary Discourse*. Reluctantly Herschel allowed his name to be used by those, including Charles Babbage, William Whewell, David Brewster, and Roderick Murchison, who saw corruption, stagnation, and unprofessionalism in Davies Gilbert's machinations to install the Duke of Sussex as his successor in the Presidency despite the Duke's having no scientific reputation. The Royal Society was not, in Herschel's view, a property to be transferred unilaterally from one owner to

---

[43] Thomas Chalmers, *The Efficacy of Prayer Consistent with the Uniformity of Nature* (London: Partridge and Oakey, 1849), p. 15.

[44] Ibid., p. 15.

another. He saw an exact parallel between the undemocratic methods of Davies Gilbert and those of the unreformed Parliament: 'if Mr Gilbert has neither the capacity to conduct the business of the Society properly, nor the spirit to hold his station in it with dignity, he ought at least not be suffered to imagine that he can hand it over like a rotten borough to any successor be his rank or station what they may, by his *ipse dixit*'.[45]

In keeping with his reformist credentials, Herschel's account of history, and the history of science, is emphatically progressive. Though he respects an immense difference between the possible scope of scientific knowledge and God's understanding of the universe, nonetheless he is optimistic about the task ahead of science:

But to ascend to the origin of things, and speculate on the creation, is not the business of the natural philosopher. An humbler field is sufficient for him in the endeavour to discover, as far as our faculties will permit, what *are* these primary qualities originally and unalterably impressed on matter, and to discover the *spirit* of the laws of nature, [or] if such a step be beyond our faculties; and the essential qualities of material agents be really *occult*, or incapable of being expressed in any form intelligible to our understandings, at least to approach as near to their comprehension as the nature of the case will allow; and devise such forms of words as shall include and *represent* the greatest possible multitude and variety of phenomena.[46]

The 'humble' task of the natural philosopher turns out to be noble in its inclusiveness and aspiration. By demarcating the separate tasks of the natural philosopher and the theologian, Herschel is limiting science's scope and at the same time strengthening its claims to speak authoritatively on the matters with which it concerns itself. And for many contemporary writers, the limits modern science placed on its scope—its abandonment of 'speculation' to religion—were a mark of science's maturity and the source of its justified self-confidence.

So although contemporary writers on science tended to favour an image of science as somewhat more bounded than that envisaged in Thomas Sprat's exuberance, this boundedness should be interpreted as a sign of growing disciplinary identity and coherence. Science, understood as a collective and depersonalized body of knowledge, had a task

---

[45] Herschel to Fitton, 18 October 1830, quoted in Jack Morrell and Arnold Thackray, *Gentlemen of Science: Early Years of the British Association for the Advancement of Science* (Oxford: Clarendon Press, 1981), pp. 53–4. In the event, Herschel was defeated in the election and refused to serve as the President of the Royal Society when the position was offered to him a decade later.

[46] Herschel, *Preliminary Discourse*, pp. 38–9.

which encompassed an immense field of study. Although most writers acknowledged that it had limits, those limits were almost infinitely far off, and did not impinge on the real authority of science except to support it by cutting it off from speculation. There remained, though, the problem of how to understand science as a part of contemporary society. Did the divisions within society have counterparts within the work or the body of science? A key question was that of the extent to which it was appropriate for different classes and sections of society to have access to scientific knowledge. And in debating this question, metaphors of landscape and boundaries were, not surprisingly, in frequent use on both sides of the argument.

Landscape metaphors had very little to contribute to the nineteenth-century debate about how to conceptualize knowledge and how to order it into a comprehensible single system. The 'Temples of Science' and 'rich domains' of learning that crop up again and again in these metaphors are too unspecific and fantastical to help in categorizing information. But they *were* useful in the parallel argument about who owned, regulated, and policed the boundaries of knowledge. In a period witnessing the birth of mass literacy and the development of a public sphere in which many kinds of print culture were widely and cheaply available, this argument was urgent and heated.

Faraday the autodidact, the Dissenting artisan, constructed landscape metaphors based on Biblical allegories about the duty of humans to cultivate and improve land, and on the picturesque's valorization of a delightful bewilderment in response to the tamed otherness of nature. Even where his metaphors indicate most anxiety about the possibility of the human mind being lost in the variety and scope of the world of learning, his chief concern nonetheless is with access to that world: he sometimes questions the safety of entering into that world but he nonetheless assumes it is his duty and pleasure to do so. Other writers, however, were using landscape metaphors to partition the world of learning so as to regulate access to it, or denying the possibility that the multitude could even perceive it to exist.

Eighteenth- and nineteenth-century writing drawing on the Protestant allegorical landscape tradition tends to be most interested in those aspects of landscape which reflect human intervention, because at bottom it is more concerned with the morality of human activity than with the wonder of the divine creation. Thus it focuses on agriculture, horticulture, and the domesticized landscape in preference to sublime, inhospitable, or wild geographies. Conservative social politics in the late eighteenth and early nineteenth centuries had easy recourse to metaphors which encoded desirable political situations as resulting from and producing desirable landscapes. As Nigel Everett puts it, during this period, 'intervention in the landscape was understood as making explicit and readable statements about the political history, the political constitution, the political future of England, and about the relations that should exist between its citizens'.[47] In literature as well as practice, landscape was saturated with political and religious meaning. In the writing of Hannah More, for instance, the development of the genre of popular political economy came together with the Protestant tradition of landscape allegory in narratives in which a set of conservative social values were expressed through images of an idealized rural landscape. Anne Stott, who successfully argues that the nuances of More's politics have been insufficiently regarded by critics hostile to her conservatism, describes the setting of More's *Village Politics* (1793) as 'both allegorical and rooted in an idealized contemporary reality'.[48] This mixed approach is characteristic of More's fiction, allowing her a great deal of flexibility in the pointedness of her narratives, and reflecting her understanding of the natural world as itself intended as a sign of a higher reality. The legibility of the natural world is a key feature in contemporary political and religious writing about landscape. For instance, More's first story for the Cheap Repository, 'The Shepherd of Salisbury Plain', opens with a lesson in the appropriate way in which to respond to landscape. The Holy Ghost, the protagonist muses, inspired Psalm 19 'as a kind of general intimation to what use we were to convert our admiration of created things; namely, that we might be led by a sight of them to raise our views from the kingdom of nature to that of grace'.[49] For More, minimal education beyond Scriptural knowledge is required for a right reading of landscape. But the great questions of early nineteenth-century education politics, who is to be educated and what knowledge is to be provided for them, demanded more complex answers, even from writers with a broadly conservative view of the problem.

[47] Nigel Everett, *The Tory View of Landscape* (New Haven, Ct.: Yale University Press, 1994), p. 7.

[48] Anne Stott, *Hannah More: The First Victorian* (Oxford: Oxford University Press, 2003), p. 140.

[49] *The Works of Hannah More*, 6 vols (London: Fisher, Fisher and Jackson, 1834), I, pp. 251–86 (p. 252).

In chapter 12 of *Biographia Literaria*, Coleridge launches an extended landscape metaphor in which he attacks the advocates of mass scientific education:

I say then, that it is neither possible or necessary for all men, or for many, to be PHILOSOPHERS. There is a *philosophic* (and inasmuch as it is actualized by an effort of freedom, an *artificial*) *consciousness*, which lies beneath or (as it were) *behind* the spontaneous consciousness natural to all reflecting beings. As the elder Romans distinguished their northern provinces into Cis-Alpine and Trans-Alpine, so may we divide all the objects of human knowledge into those on this side, and those on the other side of the spontaneous consciousness; *citra et trans conscientiam communem.* The latter is exclusively the domain of PURE philosophy, which is therefore properly entitled *transcendental,* in order to discriminate it at once, both from mere reflection and *re*-presentation on the one hand, and on the other from those flights of lawless speculation which abandoned by *all* distinct consciousness, because transgressing the bounds and purposes of our intellectual faculties, are justly condemned, as *transcendent.* The first range of hills, that encircles the scanty vale of human life, is the horizon for the majority of its inhabitants. On *its* ridges the common sun is born and departs. From *them* the stars rise, and touching *them* they vanish. By the many, even this range, the natural limit and bulwark of the vale, is but imperfectly known. Its higher ascents are too often hidden by mists and clouds from uncultivated swamps, which few have courage or curiosity to penetrate. To the multitude below these vapors appear, now as the dark haunts of terrific agents, on which none may intrude with impunity; and now all *a-glow,* with colors not their own, they are gazed at, as the splendid palaces of happiness and power. But in all ages there have been a few, who measuring and sounding the rivers of the vale at the feet of their furthest inaccessible falls have learnt, that the sources must be far higher and far inward; a few, who even in the level streams have detected elements, which neither the vale itself or the surrounding mountains contained or could supply. How and whence to these thoughts, these strong probabilities, the ascertaining vision, the intuitive knowledge, may finally supervene, can be learnt only by the fact.[50]

Coleridge puts landscape to work to justify the limited access to knowledge enjoyed by the majority of the inhabitants of this valley and of his fellow citizens in early nineteenth-century Britain. Comparing the state of educational privilege to a landscape allows Coleridge to naturalize his elitist conception of the dissemination of learning: just as there exist natural landscape boundaries on human activity, so too

[50] Samuel Taylor Coleridge, *Biographia Literaria; or, Biographical Sketches of My Literary Life and Opinions,* ed. James Engell and W. Jackson Bate, 2 vols (London: Routledge & Kegan Paul, 1983), I, pp. 236–40.

there are natural intellectual boundaries. (This analogy only works if we ignore the radical changes humans make to natural landscape boundaries by tunnelling, bridging, fording, and so on.)

Despite the Gothic unreality of the description, this naturalizing argument fits fairly well with the contemporary scientific view that the physical geography of a place determines the mentality of its inhabitants. Mary Somerville, for instance, argued that of all the external circumstances that affect 'the universal mind of a people', the most influential was 'the geographical features of the country they inhabit'.[51] Alexander von Humboldt, perhaps the most internationally-renowned scientific traveller of the period, concurred that geographical locale, including 'the peculiar influence of the configuration of the soil' profoundly influences the cultural manifestations possible and prevalent in that locale.[52]

But Coleridge goes further than the contemporary view that landscape and climate condition the kind of human society and culture produced in a given region, by making the geographical features of his imaginary landscape into a 'natural limit' to the thoughts of the inhabitants. He writes about this landscape as if it were devoid of history, making the domination of the geography over the people all the more absolute. In the landscape metaphors in Whewell's, Ritchie's, and Smith's writing discussed earlier, individuals work to match or even increase the amount of knowledge already available thanks to the activities of previous generations. But in Coleridge's metaphor there is no sense that cumulative progress is being made even by the few who seek knowledge beyond their surroundings. There are no guides to help the learner make sense of the landscape. Although the investigative few exist 'in all ages', they are isolated, gaining their information from the natural world itself rather than from their peers or predecessors. Coleridge's metaphor puts a thumb on the scales of the balance between nature and culture that usually operates in these landscape tropes. It presents the acquisition of scientific knowledge as a negotiation between the individual and the natural world, with no cumulative culture to mediate between them. Coleridge's politics of knowledge here are deeply

---

[51] Mary Somerville, *Physical Geography*, 2 vols (London: John Murray, 1848), II, p. 251.

[52] Alexander von Humboldt, *Cosmos: A Sketch of a Physical Description of the Universe*, transl. E. C. Otté *et al.*, 5 vols (London: Bohn, 1849–65), I, p. 411. One recent critic goes so far as to call Humboldt 'an environmental determinist': Aaron Sachs, 'The Ultimate "Other": Post-Colonialism and Alexander von Humboldt's Ecological Relationship with Nature', *History and Theory* 42 (2003), 111–35 (127).

conservative, rejecting the democratizing efforts of the mass education movements and insisting on a qualitative distinction between those capable of scientific enlightenment and those not.

Though landscape metaphors rarely said much about the production of knowledge, they could and did advance arguments about its ownership, as we have seen. They visualized information as a natural good, sometimes in need of cultivation (as in agricultural metaphors), sometimes already cultivated (garden metaphors). But though they imagined information in terms of landscapes which were radically remote from the industrial actuality of early nineteenth-century scientific and technical activity, they could nonetheless be used to discuss the access of the mass population to knowledge. They did this by translating urban social experience into a variety of agrarian and pastoral scenes which polarized the subtleties of early nineteenth-century politics into the clear, simple contrasts of the allegorized rural scene.

So landscape metaphors used in epistemological writing encode relationships between the individual and the world of learning as simplified, stylized versions of contemporary constructions of the relationship of humans to the natural world. They may draw on aesthetic or on economic constructions of landscape, but in either case their work is to dramatize issues of access to and ownership of knowledge by comparison with well-understood tropes of access and ownership of real estate. They rely on an enormous leap of metaphor by which culture is imaged as nature. This leap necessarily brings a range of responses to nature into play in discussions of culture; for early nineteenth-century writers these included contrasting attitudes to nature as unchanging and yet improveable, as designed by God and yet redesignable by humans, and as spectacle and sustenance.

It is not at all surprising that at the moment when mass literacy and education were becoming political topics of great urgency and constant debate, landscape metaphors were so widely and frequently resorted to in writing about knowledge. The image of the landscape allowed all shades of opinion to argue about fundamental questions of freedom, rights, the protection of humanist heritage, and the democratization of access to that heritage in terms which appealed to a very strong combination of ancient attitudes to territory and contemporary aesthetic responses to beauty.

2

Organizing the Space of Knowledge

Landscape metaphors were particularly suited to expressing the widespread feeling of optimism about scientific progress and its effect on the intellectual, moral, and material wellbeing of the nation. They could encourage a reader to visualize an immensely broad scene while also invoking a sense of ownership that has obvious connections to British imperial ventures in this period. Discovery and conquest could become conflated. Charles Daubeny's urbane *Introduction to the Atomic Theory* exemplifies the expansionist and triumphalist tendencies of this kind of rhetoric. Daubeny writes that longstanding barriers to human knowledge

have receded before the march of discovery, and [...] whole provinces, into which the ancients only penetrated by a few desultory and random incursions, have been added to the domain of modern science—real and substantial possessions, which hold about the same relation to the visionary regions of knowledge existing in the imagination of the latter, as the lands explored by Columbus bear to the fabulous Atlantis.[1]

But acquiring 'real and substantial possessions' brings with it real and substantial problems, one of which is how to organize those possessions. Making sense of the immense scope of new and newly accessible knowledge was one of the key intellectual activities of the early part of the nineteenth century. We see the problem in miniature with Mr Brooke's documents in *Middlemarch*:

'[...] I began a long while ago to collect documents. They want arranging, but when a question has struck me, I have written to somebody and got an answer. I have documents at my back. But now, how do you arrange your documents?'
'In pigeon-holes partly,' said Mr Casaubon, with rather a startled air of effort.

---

[1] Charles Daubeny, *An Introduction to the Atomic Theory* (Oxford and London: Murray, 1831), p. 3.

'Ah, pigeon-holes will not do. I have tried pigeon-holes, but everything gets mixed in pigeon-holes: I never know whether a paper is in A or Z.'

'I wish you would let me sort your papers for you, uncle,' said Dorothea. 'I would letter them all, and then make a list of subjects under each letter.'[2]

Dorothea's suggestion goes down well with Mr Casaubon, but perhaps if she had been called on to state what categories she would use for sorting the papers, her system would have seemed less straightforward. During the first half of the nineteenth century, commentators ranging from the internationally renowned to the deeply obscure published new systems for organizing the divisions of human knowledge. This multiplicity of suggestions for organizing (which generally involved hierarchizing) the branches of knowledge demonstrated in turn a multiplicity of ideological approaches to the production and transmission of information.

André-Marie Ampère, whose work on electrodynamics was a key factor in the development of that science in the 1820s and 1830s, proposed a wholesale re-classification based on a rejection of all variants of the Baconian system, which had dominated knowledge classification for centuries. Where Bacon divided the branches of knowledge up by reference to whether they belonged to the mental faculties of memory, reason or imagination, Ampère insisted that classification must be by reference not only to the objects studied by each science but also to the point of view under which they are studied. Like most of the proposed classifications of this period, Ampère's was tabulated using a tree-like structure whereby large headings were subdivided into smaller and smaller ones, culminating in a list of 128 ultimate branches of knowledge. Two basic 'règnes' divided knowledge into the cosmological sciences and the noological ones (i.e., those centred round the human understanding rather than external phenomena). After that basic division, 32 sciences of the 'first order' each yielded two of the second order, and each of those yielded two of the third order. The symmetry of the ordering was part of its rhetorical claim to authority. Literature, for example, was a first-order science among the noological ones, and it was subdivided first into bibliogy (what we might term material study of the book) and comparative literature, and these into bibliography and bibliognosy, literary criticism and philosophy of literature.[3]

---

[2] George Eliot, *Middlemarch* (1871–2) ed. David Carroll (Oxford: Clarendon Press, 1986), p. 19 (Book I, ch. 2).

[3] André-Marie Ampère, *Essai sur la philosophie des sciences, ou exposition analytique d'une classification naturelle de toutes les connaissances humaines* (Paris: Bachelier, 1834). The table listing the classifications folds out from the back of the book.

Jeremy Bentham's somewhat earlier and fairly perplexing attempt at classification, *Chrestomathia*, made a similar virtue of symmetry, relying on a system of bifurcations from the topmost category—'eudaemonics' or the study of increasing felicity—down through a series of steps based on the objects of the science, and confusingly renaming all the branches of knowledge using Greek roots. This system resulted in a separation of the sciences of material things from those of immaterial ones. For example, tracing a path down from 'eudaemonics' to what his readers knew as geometry, we pass through 'coenoscopic eudaemonics' (sciences that investigate properties common to all beings) to 'somatoscopic', and then 'pososcopic' sciences (the first investigating things having body, i.e. material things, the second investigating quantities rather than qualities), before reaching 'morphoscopic mathematics', mathematics focusing on form.[4]

Ampère's and Bentham's systems were different in many respects, most important of which was the fact that Ampère's hierarchy led from general to more and more particular classifications, where Bentham's divisions were based on judgements about the objects studied rather than the generality or otherwise of the study itself. But both writers clearly felt that the ordering of the branches of knowledge ought to be done according to a logical, regular, undeviating system, of which symmetry was one of the desiderata and a mark of success, though Ampère admitted that his system would have to be modified as new connections between facts became known; his classification was subordinate to the present state of knowledge, not vice versa.[5]

These were just two of a great many proposals published in the early nineteenth century for reforming the classification of the branches of learning, an endeavour which was often said to be vital for both the advancement of knowledge and for bringing students and general readers up to date in a given field. But the majority of dictionaries and encyclopaedias of arts and sciences continued to arrange their material using what one encyclopaedist called the 'unphilosophical, inconvenient' principle of alphabetical order.[6] And the classification of

---

[4] Jeremy Bentham, *Chrestomathia: Being a Collection of Papers, Explanatory of the Design of an Institution, Proposed to be Set on Foot, Under the Name of The Chrestomathic Day School [...] for the Use of the Middling and Higher Ranks of Life* (London: Payne and Foss, 1816). Table V.

[5] Ampère, *Essai sur la philosophie des sciences*, p. 10.

[6] Edward Smedley, letter to Charles Babbage, 17 July 1826, quoted in Richard Yeo, *Encyclopaedic Visions: Scientific Dictionaries and Enlightenment Culture* (Cambridge: CUP, 2001), p. 252.

what we would call 'disciplines' was only an instance, though a very visible and contentious one, of the much wider problem of presenting the enormous range of knowledge to a new, fragmented, and increasingly self-aware readership. In the next chapter I will return to the problem of disciplinization and the splitting-off and joining together of sciences. But first we need to consider the question of spatial discourse in less formal kinds of organization of knowledge.

Among the most important purposes to which spatial discourse was put in the literary and scientific writing of the early nineteenth century was the organization of knowledge into accessible pieces and comprehensible patterns. Writers used metaphors of space to describe knowledge as a kind of field or landscape through which learners and readers travelled, as we saw in the last chapter; and a more abstract kind of metaphor could be used which envisaged knowledge as an array of items to be organized into shapes which would impose order and meaning on their apparent randomness. This kind of metaphor involves a more complex relationship between the items of knowledge and the observer or possessor of those items. Two examples of these metaphors will illustrate some of the issues involved.

The early nineteenth century saw two contradictory movements in the organization of knowledge—on the one hand, there was a clear shift towards increasing specialization and insulation of one area of learning from the others, and on the other hand there were significant efforts towards unifying and harmonizing these different specialisms. The principles by which knowledge could be categorized were deeply contested. Early nineteenth-century encyclopaedias give interesting examples of rival methods of organizing knowledge, particularly since one of their chief aims is to make knowledge accessible as well as simply stable. Encyclopaedias in this period frequently began by giving fairly lengthy accounts of the method or system they used for achieving these aims. The 1842 edition of the *Encyclopaedia Britannica* devotes several pages of its Preface to a history of previous encyclopaedic methods, concluding with a denunciation of Diderot's *Encyclopédie* with its 'irreligious and revolutionary designs'.[7] It was common practice in writing encyclopaedias to use spatial metaphors to explain the method of organization that governed the information that was to follow. Richard Yeo has

---

[7] *Encyclopedia Britannica*, 7th edn (1842), I, p. xii.

shown how metaphors based on maps, for example, were repeatedly used in these explanations.[8] Map metaphors, by making an analogy between knowledge and property, open crucial rhetorical fields allowing discussion of ownership, access, and legitimacy. Coleridge was thus making a disingenuous claim when he wrote in the first section of the *Encyclopedia Metropolitana* that 'to *methodize* such a compendium, has either never been attempted, or the attempt has failed, from the total disregard of those general connecting principles, on which Method essentially depends'.[9] On the contrary, he was writing in a long tradition of encyclopaedists when he used spatial metaphors to describe the project, its method, and the fields of knowledge it aimed to methodize, as a 'labyrinth', a 'path or way of transit', and a 'whole circle'.[10]

All spatial metaphors that perform organizational work inscribe power relationships into the spaces they arrange; these relationships are what constitute organization. In its most basic sense, this means that when used for this purpose, spatial metaphors make one kind of space (or the things in it) subservient to another. At a more sophisticated level, spatial metaphors can govern the power relationships between the spaces they describe and the reader. The two metaphors, or models, I focus on in this chapter are examples of two quite different understandings of power relationships and hence organization of knowledge. The hub-and-ray model of spatial arrangement organizes objects (or ideas, or information) as a pattern of circles and straight lines, like a wheel whose spokes all lead to a central point. In comparison, the aerial view model organizes by flattening an array of objects into a plane surface and lifting the observer above that plane so that it is all visible from a single point of view. The aerial view's power relationships are derived from hierarchizing the vertical levels it imagines, whereas the hub-and-ray model's relationships are based on interiority and proximity to a central point.

Both models are familiar to present-day as well as to early nineteenth-century readers. In critical theory, the hub-and-ray arrangement is

[8] For a shorter version of Yeo's argument, see 'Reading Encyclopedias: Science and the Organization of Knowledge in British Dictionaries of Arts and Sciences, 1730–1850', *Isis*, 82 (1991), 24–49. *Encyclopaedic Visions* is the full-scale version.

[9] Samuel Taylor Coleridge, 'Preliminary Treatise on Method', *Encyclopedia Metropolitana*, 5th edn (1851), p. 13.

[10] Ibid., pp. 13, 15, 14.

the basis of the centre-and-periphery trope as well as of that recently ubiquitous epistemological design, the panopticon. The aerial view is yet more familiar to us in the form of maps and plans. These models, and others like them, are very powerful tools partly because they belong both to the physical and the mental world, existing as concrete things (panopticons were actually built; maps were and are actually printed and used) as well as abstractions in both nineteenth-century and contemporary culture. They were, and are, good ways to explain how each one of a set of things relates to every other one. But part of what makes them good tools is the way in which they structure not only the relationships between the objects in the set (such as cells in the panoptical prison, or cities on the map) but also the relationships between the model itself and the person reading it. These models are comprehensible because of our experience of sensory objects in the physical world—things we have seen (such as wheels) and ways in which we have seen them (such as from above).

Because they respond to sensory experience they are subject to history, becoming imbricated in changing aesthetics, for example, and reacting with particular sensitivity to changes in technologies of vision. But both models try to expunge history from their manifestations. Unlike the metaphors discussed in the last chapter, i.e. tropes that envisage knowledge as a landscape to be traversed and experienced individually, these models attempt to remove the randomness from the encounter with knowledge. They try to do away with the bewilderment of the lost wanderer in the realms of knowledge, as well as the surprised pleasure of the picturesque tourist, and all such responses involving uncertainty or incomprehension. Their emphasis on pattern rather than improvization results in claims to the authority of distance rather than that of intimacy, and to atemporality. Both the hub-and-ray model and the aerial view seek to offer an escape from history, particularly by providing a method of non-sequential reading. But while the hub-and-ray model regards order, arrangement, as existing *before* the objects ordered or arranged, the aerial view tries to impose arrangement on an already existing set of objects. The difference between arrangement and objects in these models might be thought of as analogous to that between space and place—the one general and transferrable, the other specific and fixed. Spatial organizations of knowledge prioritize one or the other, and with this prioritization they also determine the relationship of the reader/observer to the knowledge observed.

HUBS AND RAYS

Coleridge's image in the *Encyclopedia Metropolitana* of a 'whole circle' as a representation of the arts and sciences was characteristic of encyclopaedists in its use of spatial metaphor, but also an example of one of his own favourite spatial tropes. Elsewhere in his 'Treatise on Method' he uses the image of a circle to discuss the key concept of what constitutes an idea. To a geometrician, he writes, a circle is 'a clear, distinct, definite' idea, but a boy only knows 'that his hoop is round', though this knowledge will eventually help him learn that 'in a circle, all the lines drawn from the centre to the circumference are equal'.[11] Coleridge's point is that learning proceeds from vagueness to clarity, but the image of a circle—particularly a circle with its radii visible or marked—as an example of perfect knowledge is one he returns to frequently, and one that allows him to emphasize the fundamental unity of all true knowledge. In *Biographia Literaria*, for instance, he uses this same spatial model to envisage the final unification of all competing systems of philosophy. Coleridge agrees with Leibniz (and perhaps with *Middlemarch*'s Casaubon) that 'the criterion of a true philosophy' is that it will 'at once explain and collect the fragments of truth scattered through systems apparently the most incongruous'. All the particles of truth found in all the schools (which Coleridge lists in a seven-item catalogue grossly burdened with proper names and abstract nouns) will be 'united in one perspective central point, which shows regularity and a coincidence of all the parts in the very object, which from every other point of view must appear confused and distorted'.[12] The image is of lines radiating in towards a centre from which all truth will appear ordered. De Quincey's third essay on 'Style' similarly uses a hub-and-ray image to describe the organization of apparently random information (the uneven distribution of the stars in the night sky) as having pattern and order invisible to humans but consolingly imaginable:

It is thought by some people that all those stars which you see glittering so restlessly on a keen frosty night in a high latitude [...] are in fact all gathered into zones or strata; that our own wicked little earth (with the whole of our peculiar solar system) is a part of such a zone, and that all this perfect geometry

---

[11] Ibid., p. 17.   [12] Coleridge, *Biographia Literaria*, I, p. 247.

of the heavens, these radii in the mighty wheel, would become apparent if we, the spectators, could but survey it from the true centre,—which centre may be far too distant for any vision of man, naked or armed, to reach.[13]

Both De Quincey's and Coleridge's metaphors structure information as arrayed in a circular pattern around a central point at which the observer stands. This structure is imagined as powerful enough to make sense of the fragmented and disparate bits of information it contains. I will call this the 'hub-and-ray' structure, and argue that where it was used in early nineteenth-century writing it had a subtle but profound effect on that text's politics of knowledge.

The hub-and-ray images we have seen so far are used to make random knowledge structured and hence accessible, but accessible in a quasi-mystical, visionary way. Compared with the practicalities of organizing information in encyclopaedias, for instance, this is a distinctly emotional way of dealing with the organization of knowledge, tending to expression in heightened, sometimes even almost religious rhetoric. During the *Naturphilosophie* movement, for example, the hub-and-ray model was used by the German Romantic writer Novalis to express a conception of the unity of the human mind, the natural world, and ways of understanding that world. In his fable *The Novices of Sais* Novalis figures knowledge as a collection of miscellaneous natural objects including seashells, stones, flowers, *objets trouvés*, which are arranged in rows by a master. The climactic moment of the narrative is the joyous unification of these rows into a pattern forming a single perspective point like the one Coleridge hopes will unite all philosophical truths:

Soon the novice stepped into our midst with ineffable joy in his face; he was carrying a humble little stone, of a strange shape. The teacher took it in his hand and kissed it a long long while, then he looked at us with tears in his eyes and laid the little stone in an empty space among the other stones, where many rows came together like spokes.

Never shall I forget those moments. It was as though our souls had known a bright and fugitive presentiment of this wondrous world.[14]

The mere presence of the final stone is enough to make a significant pattern of the previous arrangement. In all hub-and-ray models, including

---

[13] Thomas De Quincey, 'Style III' (1840), in *The Works of Thomas De Quincey*, gen. ed. Grevel Lindop, 21 vols (London: Pickering and Chatto, 2000–3), 12 (2001), pp. 45–63 (p. 57).

[14] Novalis, *The Novices of Sais*, trans. Ralph Manheim (New York: Valentin, 1949), pp. 11–13.

this one, location is symbolic. Here, the space at the centre of the radial rows pre-exists the stone which fills it: the meaning of the stone, or the function it will fulfil, is predetermined by the spatial characteristics of the context.

Novalis's stones and Coleridge's competing philosophical systems are both collections of disparate items which add up to more than the sum of their parts because of the arrangement in which they occupy space. Timothy Mitchell's phrase 'the world-as-exhibition' may be helpful here.[15] The notion of 'world-as-exhibition' is derived from Heidegger's conception of a 'world picture', that is, a method, supposedly characteristic of modernity, of understanding the world as a picture.[16] The 'world-as-exhibition' model is intended to be nonlinguistic: items arranged as an exhibition need only to be seen, not read, to be understood. This is an aspiration shared also by both hub-and-ray and aerial view forms of spatial arrangement of information. Like the hub-and-ray model, the idea of the world-as-exhibition asserts an order that pre-exists the objects it orders. Derek Gregory notes that 'one of the central pinions of the world-as-exhibition was a conception of order that was produced by—and resided in—a structure that was supposed to be somehow separate from what it structured: A framework that seemed to precede and exist apart from the objects that it enframed'.[17] Novalis's image of a central space waiting for the stone which will fill it is clearly an instance of this structural precedence. The modern discipline of geography depends heavily on various forms of this conception of a preceding structure. Peter Haggett's influential geography of locational analysis, for instance, centres on a similar conception of order, one which 'depends not on the geometry of the object we see, but on the organizational framework in which we place it'.[18] Gregory describes the geographical practice which stems from this conception of geographical theory as 'the decomposition of a regional system into a series of abstract geometries: movements, networks, nodes,

---

[15] Derek Gregory, *Geographical Imaginations* (Oxford: Basil Blackwell, 1994), part of title of chapter 1. The term 'world-as-exhibition' is used by Timothy Mitchell in *Colonising Egypt* (Cambridge: CUP, 1988), pp. 13–14.

[16] Martin Heidegger, 'The Age of the World Picture', in *The Question Concerning Technology, and other Essays*, trans. William Lovitt (New York: Harper and Row, 1977), pp. 115–54 (p. 129): 'world picture, when understood essentially, does not mean a picture of the world but the world conceived and grasped as picture'.

[17] Gregory, *Geographical Imaginations*, p. 53.

[18] Peter Haggett, *Locational Analysis in Human Geography* (London: Arnold, 1965), p. 2; quoted in Gregory, *Geographical Imaginations*, p. 53.

hierarchies and surfaces'.[19] A well-known example appears in Edward Soja's account of Los Angeles as needing to be 'reduced to a more familiar and localized geometry to be seen'.[20] In other words, these geographies transform place into space: physical landscape becomes geometric abstraction.

An exhibition is a collective unit, comprising a number of individual items, no single one of which can be said to define the exhibition as a whole. These items, such as the ancient philosophies in Coleridge's list, and Novalis's 'stones, flowers, insects of all sorts', may be individually of little importance, but by their insertion into a spatial framework they gain significance.[21] Witnessing the way that the central perspective point makes sense of these exhibitions is the spectator. It is the spectator, rather than a strangely shaped stone or an as-yet-unknown unifying philosophy, that provides the conceptual significance of the exhibited objects.

Here a connection may be suggested with the familiar image of the panopticon, which can be thought of as a central hub from which radiating sight lines reach their termination in a circular exhibition. The purpose of the panopticon is to maximize the power of the spectator at the center of the spokes. However, if the hub-and-ray and panopticon models are regarded as designs, or visual patterns, it is clear that there is an important difference between them: the circular boundary joining the rays at their extremes, which the panopticon requires, is missing from the hub-and-ray arrangement. In principle the rays can extend out from the centre indefinitely. This difference is important for the meanings, as well as the appearances, of the two models. The panopticon's outer circle provides a sort of closure, whereas there is no defined limit to the extension of the rays in the other image. Novalis does not tell us how long the rows of stones and flowers are: theoretically they might extend to infinity and organize all conceivable space. The panopticon, on the other hand, can serve no purpose unless there is a limit to its sight lines. This element of closure essential to the panopticon creates the possibility that the model can be multiplied. It is hard to imagine how more than one hub-and-ray arrangement could be accommodated in the same region of space: in both Coleridge's and Novalis's versions of this arrangement, only one hub appears; only one is required. But with

[19] Gregory, *Geographical Imaginations*, p. 53.

[20] Edward W. Soja, *Postmodern Geographies: The Reassertion of Space in Critical Social Theory* (London: Verso, 1989), p. 224, quoted in Gregory, *Geographical Imaginations*, p. 301.

[21] Novalis, *Novices of Sais*, p. 7.

the panopticon arrangement, multiples are possible. The recognition that the panopticon terminates in a boundary prompts the question of what space exists outside that boundary, and how it is to be ordered. In the classic Benthamite design for a panopticon, interior gratings facing towards the hub allow the unseen central spectator to look into the cells, but there are also exterior windows, facing out of the panopticon, to allow light into the cells.[22] The panopticon is therefore not a sealed unit. The presence of space outside the panopticon is recognized, and access (visual, of course, not physical) to that outside space is permitted. The hub-and-ray arrangement potentially organizes all space, because it has no boundary; the panopticon, being self-contained, draws attention to the space outside its perimeter.

If the outer perimeter of each panoptical unit is circular, the multiples will not tessellate: empty space will necessarily be left between them. Only if the perimeter forms a square or a hexagon, for instance, will the multiples be able to exclude non-panoptical space. Bentham was aware of this problem, and suggested combining four 'rotundas' to form a circle, square, or rectangle, according 'to the prevailing ideas of beauty or local convenience'.[23] The fact that circles will not tesselate was used in the middle ages as an argument against the plurality of worlds: worlds were conceived as spherical, and since spheres (unlike cubes) can touch only at points, not by surfaces, a postulated universe containing more than one spherical world would necessarily also include some 'waste' space between those worlds. That space would have to be void, since filled space belongs to a world; and the possibility of void space was denied because it seemed to imply limitations to the power and presence of God. Since there could be no void space in the interstices between worlds, there could be no more than one world.[24] In both the case of the circular panopticon and that of the spherical world, unorganized space between the non-tessellating shapes represents a threat to the

---

[22] See Bentham's plan of the panopticon, in *Panopticon; or, The Inspection-House* [...] (London: Payne, 1791; repr. in *The Works of Jeremy Bentham*, ed. John Bowring, 11 vols (Edinburgh: Tait, 1843)), IV, pp. 37–172 (pp. 40–1). See also Foucault's description of the panopticon in 'The Eye of Power', in *Power/Knowledge: Selected Interviews and Other Writings, 1972–1977*, ed. Colin Gordon, trans. Colin Gordon *et al.* (Brighton: Harvester, 1980), pp. 146–65 (p. 147).

[23] Bentham, *Works*, IV, p. 44.

[24] See Pierre Duhem, *Medieval Cosmology: Theories of Infinity, Place, Time, Void and the Plurality of Worlds*, ed. and trans. Roger Ariew (Chicago: University of Chicago Press, 1985), pp. 387–8 for a fuller discussion of these arguments as they appear in thirteenth-century Scholasticism.

organization of the space within the circle or sphere. Despite Foucault's suggestion that the ordering of the panopticon is so powerful that even the central observer or spectator is trapped, or disciplined, within its spatial logic, perhaps the hub-and-ray model is more powerful as a spatial model for conceptualization than the closed panopticon.[25] The hub-and-ray model's possible extension to infinity makes it singular rather than multiple, and allows it to organize all space, rather than just the space of its occupation.

Despite these differences between hub-and-ray models and the panopticon, though, it is important to remember that in both cases it is the underlying spatial organization, rather than the objects (or people) exhibited, which determines the comprehensibility of the exhibition. As De Quincey put it in his suggestion that the night sky might be organized with a hub-and-ray model, 'it is most instructive to see how many apparent scenes of confusion break up into orderly arrangement when you are able to supply an *a priori* principle of organization to their seeming chaos.'[26]

Although Foucault builds on Bentham's point that the hub of the panopticon would not necessarily have to be staffed, because eventually surveillance would come to be assumed rather than perceived by the inhabitants of the outer circle, the spectator or observer is nonetheless crucial to the meaning of the spatial structure.[27] The physical presence of a human spectator may, at any given moment, be in doubt, but for the purposes of the prisoner or the exhibit, it is to be assumed. The spectator is sometimes virtual, but is no less powerful for that virtuality. The hub-and-ray model, incorporated in both the panopticon and the 'central perspective point' of Coleridge's image, represents in itself an oddly powerful coming together of ideas about vision and the visionary.

This emphasis on spectatorship and modes of vision means that the hub-and-ray model could be used for commenting on accessibility—in the context of epistemological writing, accessibility of knowledge—i.e. for asking who can see what kinds of information. Again in *Biographia Literaria*, Coleridge uses the hub-and-ray model as he refines the elitist

[25] Foucault, 'The Eye of Power', p. 156: the panopticon is 'a machine in which everyone is caught, those who exercise power just as much as those over whom it is exercised'.

[26] De Quincey, 'Style III', p. 57.

[27] Foucault, 'The Eye of Power', p. 155: 'an inspecting gaze, a gaze which each individual under its weight will end by interiorising to the point that he is his own overseer, each individual thus exercising this surveillance over, and against, himself'.

conception of science and philosophy which he seemed to be offering in the long landscape metaphor quoted in my previous chapter. Here Coleridge does not limit philosophy purely to an educated class, but rather to a meritocratic elite which includes 'natural' philosophers:

> Whoever is acquainted with the history of philosophy, during the two or three last centuries, cannot but admit, that there appears to have existed a sort of secret and tacit compact among the learned, not to pass beyond a certain limit in speculative science. The privilege of free thought, so highly extolled, has at no time been held valid in actual practice, except within this limit; and not a single stride beyond it has ever been ventured without bringing obloquy on the transgressor. The few men of genius among the learned class, who actually did overstep this boundary, anxiously avoided the appearance of having so done. Therefore the true depth of science, and the penetration to the inmost centre, from which all the lines of knowledge diverge to their ever distant circumference, was abandoned to the illiterate and the simple, whom unstilled yearning, and an original ebulliency of spirit, had urged to the investigation of the indwelling and living ground of all things.[28]

In this passage, lines are made to stretch to infinity (their circumference is 'ever distant') and suggest extension in time, through the history of philosophy as well as through its different kinds. This hub-and-ray model is the culmination of a series of linear spatial images ('limit', 'boundary'), that together form the kind of disciplinary landscape discussed at more length in the previous chapter. Coleridge's image of a circle is a reorganization and aggrandisement, rather than a contradiction, of the linear images which precede it: the imagery, though not the argument, of the text realigns these vectors into the hub-and-ray image of the climax. The narrative trajectory is that of a jumble of lines being organized into a meaningful structure. The role of the spectator differs from that in Novalis's version of the image, for instance, but the determining and informing effect of the central space remains.

After this passage, Coleridge proceeds, in high oratorical rhetoric, to combine the imagery of *Naturphilosophie* and the Bible into a plea for a more understanding reception of the 'uneducated man of genius': 'speculative science' is reminded that it must make space for those 'branded as fanatics and fantasts', whose mode of seeing is visionary.[29] The educated observer perceives items of knowledge in a different way from the uneducated observer, according to Coleridge.

---

[28] Coleridge, *Biographia Literaria*, I, pp. 147–8.
[29] Ibid., I, pp. 150, 148, 149.

Spatial models for the organization of knowledge provide roles for observers, and crucial to those roles is the determination of different methods of seeing. Coleridge uses the hub-and-ray metaphor to suggest very long, perhaps infinite, sight lines, and associates them with visionary seeing. Wordsworth, in the aerial view organizational image I discuss next, provides a more definite and unequivocal position for the observer, and extends sight lines not away to the ever-distant horizon, but downwards, to a definite, describable plane below. In so doing, Wordsworth figures the information organized as prior to the organizational structure he offers, unlike Coleridge's or Novalis's hub-and-ray images, where the items organized are arbitrary and almost irrelevant, since the power is in the arrangement itself, not the particular information it structures. This reversed anteriority provides Wordsworth's aerial view image with a different temporal structure, and different patterns of authority.

AERIAL VIEWS

The first descriptive section of Wordsworth's *Guide through the Lakes* begins, not with an account of Lake scenery, nor indeed of actual scenery at all, but with a report of a model:

At Lucerne, in Switzerland, is shewn a Model of the Alpine country which encompasses the Lake of the four Cantons. The Spectator ascends a little platform, and sees mountains, lakes, glaciers, rivers, woods, waterfalls, and vallies, with their cottages, and every other object contained in them, lying at his feet; all things being represented in their appropriate colours. It may be easily conceived that this exhibition affords an exquisite delight to the imagination, tempting it to wander at will from valley to valley, from mountain to mountain, through the deepest recesses of the Alps. But it supplies also a more substantial pleasure: for the sublime and beautiful region, with all its hidden treasures, and their bearings and relations to each other, is thereby comprehended and understood at once.[30]

---

[30] William Wordsworth, *A Guide through the District of the Lakes in the North of England, with a Description of the Scenery, &c. for the use of Tourists and Residents*, 5th edn (Kendal: Hudson and Nicholson; London: Longman and others, 1835), in *The Prose Works of William Wordsworth*, ed. W. J. B. Owen and Jane Worthington Smyser, 3 vols (Oxford: Clarendon Press, 1974), II, pp. 121–465 (p. 170).

It is clear that this is a description of a tangible spatial metaphor, a map which requires minimal interpretation. Despite the apparent unity and hermeneutic self-sufficiency of the model, it relies, in Wordsworth's description, on the observer, whose importance is signalled by the capitalization of 'Spectator' in the second sentence. As in the characteristic colonizing standpoint, the 'monarch-of-all-I-survey' trope identified by Mary Louise Pratt, the observer is placed on an eminence from which to command the landscape.[31] This manoeuvre is also reminiscent of the surveyor's habitual ascension of high places, necessary to obtain triangulation for cartographic understanding of the landscape.[32] But it is a standpoint available for domestic consumption of the landscape as well as public surveying of it. Wordsworth complains about newcomers to the Lake District who refuse to build their houses after the model of the old dwellings and instead build them in exposed positions on the summits of hills. He attributes this to an inappropriate and insensitive 'craving for prospect'.[33] Their houses express a relationship of dominance over, rather than integration with, the landscape; they had learnt to crave for prospects, presumably, by absorbing the conventions of sublime art and writing.

Wordsworth adopts a sort of imperial gaze of his own in the opening of the *Guide*. One expression of it is the metaphor of mastery which lies uneasily alongside literal physical description in the phrase 'lying at his feet'. This phrase occurs again in Wordsworth's poem 'View from the Top of Black Comb', published in 1815, and this time it is more clearly connected with imperial pride. From the summit of the mountain, described as 'the imperial station', Wordsworth looks down on Mona's Isle which

---

[31] Mary Louise Pratt, *Imperial Eyes: Travel Writing and Transculturation* (London: Routledge, 1992), p. 201. Pratt's account of the monarch-of-all-I-survey trope is strongly associated with but not necessarily characterized by the raised viewpoint of the observer. It is, nonetheless, in significant ways, a spatialization: 'the monarch-of-all-I-survey scene, then, would seem to involve particularly explicit interaction between esthetics and ideology, in what one might call a rhetoric of presence' (p. 205).

[32] The deconstructionist cartographer J. B. Harley plays reflexively on this source of the representational power of a high viewpoint: 'across this broad conceptual landscape I shall pinpoint three eminences from which to trace some of the more specific ideological contours of maps': 'Maps, Knowledge, and Power', in Denis Cosgrove and Stephen Daniels, eds, *The Iconography of Landscape: Essays on the Symbolic Representation, Design and Use of Past Environments* (Cambridge: CUP, 1988), pp. 277–303 (p. 278).

[33] Wordsworth, *Guide Through the Lakes*, p. 211.

> now appears
> A dwindled object, and submits to lie
> At the Spectator's feet. [...][34]

The idea of a landscape submitting to the spectator who observes from high ground is inflected rather differently in Wordsworth's description of the model at Lucerne. While Black Comb offers a view of colonial dominions (Scotland, Wales, and Ireland are mentioned, within framing references to Britain at the beginning and end of the poem), the model at Lucerne offers a view coloured by the emerging quasi-colonialism of the tourist trade. In both examples, though, the height from which the observer looks at the landscape below him is the measure of his mastery of it.

In the late twentieth century, Michel de Certeau described the experience of a person viewing New York from the top of the World Trade Center in similar terms:

His elevation transfigures him into a voyeur. It puts him at a distance. It transforms the bewitching world by which one was 'possessed' into a text that lies before one's eyes. It allows one to read it, to be a solar Eye, looking down like a god.[35]

Distance from the scene observed creates a sense of authority in the observer, an authority which may be that of a 'god' or a 'voyeur', or in this case, both at once. Certeau's relevance to Wordsworth's Lucerne model is in his emphasis on the textuality, the readability, of a landscape seen from a distance. The essence of this readability is, for Certeau, spatial. He sees people walking in the city streets as writing an urban text which they cannot read because they are immersed in it.[36] The observer on top of a skyscraper, on the other hand, can read the city because he is far enough away from it to be able to see all of it at once. Derek Gregory discusses this aspect of Certeau's work as an important corrective to the writing of theorists such as Edward Soja, whose study of the geography of Los Angeles, Gregory argues, excludes the perspective of the inhabitants of the city.[37] Certeau, by contrast, who adopts the elevated view point, as Derek Gregory puts it, '*only to vacate it at the*

---

[34] William Wordsworth, 'View from the Top of Black Comb', in *Shorter Poems, 1807–1820*, ed. Carl H. Ketcham (Ithaca, N.Y.: Cornell University Press, 1989), pp. 99–100, l. 21.

[35] Michel de Certeau, *The Practice of Everyday Life*, trans. by Steven Rendall (Berkeley: University of California Press, 1984), p. 92.

[36] Ibid., p. 93.   [37] Gregory, *Geographical Imaginations*, p. 301.

first opportunity', distinguishes between vision and knowledge and thus gives authority back to the city-dwellers who know their city even though they cannot see the whole of it.[38] Feminist and postcolonialist critiques of the privileged view from above and the kind of knowledge it implies suggest that the desire to see all of a scene at once indicates an inappropriate power relationship of observer to observed. Donna Haraway, notably, calls the view from above 'the god-trick of seeing everything from nowhere' and urges instead that 'there is good reason to believe vision is better from below the brilliant space platforms of the powerful'.[39]

Wordsworth's pleasurable gaze from a high place is not greatly concerned with democratizing vision. The distance of the spectator from the real landscape in the Lucerne model is registered in Wordsworth's vocabulary of display. The landscape's 'hidden treasures' are brought to light by this model, which is actually described as an 'exhibition'. The aerial view combines the dominance of the observer with the display of the observed, resulting in a quasi-erotic aesthetics. This quasi-eroticism means that the 'treasures' of the Alps are at once seen and hidden. But the objects viewed from above cannot in themselves attempt seduction of the observer: the relationship of dominance and submission must not be destabilized, and is enforced by the fact that the model attempts a complete and literal realistic representation of the landscape. The objects are shown in their 'appropriate colours', a phrase which carries a suggestion of 'in their true colours', a further guarantee of authenticity. Nonetheless, the Lucerne model's claims to realism are undermined by the self-conscious conversion of the elements of the physical world into exhibits. To argue this is to recognize the change in the relationship between observer and landscape produced by an increase in remoteness of the one from the other, an effect produced by the smallness of the scale of the model valleys and buildings, together with the observer's distance from the surface of the model. Realism is undermined by the distance of the observer from the observed, because this distance inscribes mythical power relationships between the full-size tourist and the miniaturized landscape. This is one reason why theorists like Haraway propose that local knowledge is ultimately more reliable, because less distant, than global overview knowledge.

---

[38] Ibid., p. 301.
[39] Donna Haraway, *Simians, Cyborgs and Women: The Reinvention of Nature* (London, Free Association Books, 1991), p. 189, pp. 190–1.

In Wordsworth's report, this Alpine mock-landscape is entirely without inhabitants: the only human presence is that of the Spectator. Although human habitations are marked, the landscape is wholly open to the acquisitive eye of the onlooker. In this way, it resembles other spatial discourses including, to use Mary Louise Pratt's term, the 'European discourse of landscape' which 'deterritorializes indigenous peoples, separating them off from territories they may once have dominated, and in which they continue to make their lives'.[40] This point about the absence of people from colonial and related discourses has been central to recent scholarship on cartography and landscape description. Geoff King cites the example of sixteenth- and seventeenth-century maps of North America, noting that 'the presence of the indigenous peoples on the land as any kind of active subjects is usually denied in frontier discourse, cartographic or otherwise'.[41] More generally, Brian Harley comments that

Maps as an impersonal type of knowledge tend to 'desocialise' the territory they represent. They foster the notion of a socially empty space. The abstract quality of the map, embodied as much in the lines of a fifteenth-century Ptolemaic projection as in the contemporary images of computer cartography, lessens the burden of conscience about people in the landscape. Decisions about the exercise of power are removed from the realm of immediate face-to-face contacts.[42]

The Lucerne model effects a similar removal from the realm of human contact. This is uncharacteristic of Wordsworth's approach in this text: Jonathan Bate's judgement that 'in the *Guide* [...] people are seen firmly in relation to their material environment' is more typical of the volume overall.[43] Nonetheless, it is interesting that in his graphic simplification of the environment, first of the Alps, then (as we shall

---

[40] Pratt, *Imperial Eyes*, p. 135.

[41] Geoff King, *Mapping Reality: An Exploration of Cultural Cartographies* (Basingstoke: Macmillan, 1996), p. 104.

[42] Harley, 'Maps, Knowledge, and Power', p. 303. Harley goes no further back in cartographical history than fifteenth-century mapping. Prior to the Renaissance adoption of linear perspective, maps necessarily relied far less on the aerial view. For a brief but suggestive discussion of perspective, drawing on Denis Cosgrove's conception of Renaissance linear perspective as essentially humanist, see Gregory, *Geographical Imaginations*, pp. 389–91. For Cosgrove's account of perspective see 'Prospect, Perspective and the Evolution of the Landscape Idea', *Transactions of the Institute of British Geographers*, 10 (1985), 45–62.

[43] Jonathan Bate, *Romantic Ecology: Wordsworth and the Environmental Tradition* (London: Routledge, 1991), p. 46.

see later) of the Lakes, Wordsworth follows a model based on geology and ecology, rather than on human geography. Unlike the 'landscape of knowledge' trope discussed in my previous chapter, the aerial view and hub-and-ray images tend to erase the human from their ordering of information in the same way that they tend to erase the historical. These erasures are a necessary part of their attempt to produce an alternative kind of legibility.

The Lucerne diorama offers two distinct methods of reading, which Wordsworth divides into two sentences. His first way of reading this mock landscape is simply a modified version of the way one might 'read', or experience, the real landscape: that is, by travelling through it, either by the gaze, or by foot. Wordsworth describes the observer as feeling tempted to experience the landscape in a haphazard fashion; his sentence lists and repeats landscape features: 'from valley to valley, from mountain to mountain'. The repetitions emphasize the fact that this traversing will take time. But the second method of reading which this Wordsworth passage offers is more striking, and though still temporal as well as spatial, it is temporal in a more dramatic and complicated way.

This method of reading constitutes the 'more substantial pleasure' of the Lucerne model, and is best examined by comparison with two Scriptural passages from which Wordsworth's description draws its strength. In the account of Christ's temptation by Satan in Matthew's gospel, Christ is forced to observe the world from a height, and is asked to read the landscape below him:

Again, the devil taketh him up into an exceeding high mountain, and sheweth him all the kingdoms of the world, and the glory of them;
And saith unto him, All these things will I give thee, if thou wilt fall down and worship me.[44]

Wordsworth's description in the *Guide* bears a resemblance to Matthew's account, particularly given Wordsworth's use of the word 'tempting'. But a comparison with Luke's version of the same episode makes the spring of Wordsworth's 'more substantial pleasure' clear:

And the devil, taking him up into an high mountain, shewed unto him all the kingdoms of the world in a moment of time.[45]

From a height one seems to see everything instantaneously. It is as though distance removes one not only from responsibility, but also

[44] Matthew 4.8–9.   [45] Luke 4.5.

from the passage of time. Instantaneous knowledge, acquired visually and apparently without need of interpretative codes, constitutes a revelation, or, to use a contested term of this period, an 'intuition', and this is what the Swiss model landscape gives. In this way, the Gospel echoes combine with the remoteness from day-to-day existence to give the model Alpine landscape a visionary quality, with the Spectator figuring as divine or inspired, and able to take in information through the spatial ordering of the objects without any time passing: 'all its hidden treasures, and their bearings and relations to each other, [are] thereby *comprehended and understood at once*'. Wordsworth is trying to use a spatial model for organizing knowledge as a way of short-cutting or indeed by-passing the process of reading the information of the landscape. Reading as we normally understand it must be a temporal process: it cannot happen all at once. Transforming the gathering of information into an instantaneous act alters the 'reader'ʼs relationship with that understanding. As well as aggrandizing the spectator or reader by pretending to give him a godlike epistemological experience, it suggests that the information being delivered is congruent with itself, that it contains no mismatches, exceptions, or holes. It also suggests that that information is essentially unchanging: by eliding the temporality of experiencing the landscape, Wordsworth elides the temporal aspect of the landscape itself.

A number of critics have compared Wordsworth's book with those of other commentators on Lakeland scenery, particularly Thomas West, whose *Guide to the Lakes* was published in 1778, and republished, with addenda by other writers, in a number of editions during the late eighteenth century. Don Gifford comments that West's popular *Guide*, and its imitators, 'encouraged the tourist to experience landscape not as a continuum though which he moved but as a sequence of static compositions'. In contrast, however, 'at much the same time, Wordsworth and Coleridge were turning away from the reliance on fixed prospects and stations towards a more fluid experience of the natural world, itself perceived as organism in motion'.[46] The opening

---

[46] Don Gifford, *The Farther Shore: A Natural History of Perception, 1798–1984* (London: Faber and Faber, 1990), p. 20. John Wyatt's more recent *Wordsworth and the Geologists* sees a rather similar and very pronounced difference between Wordsworth's and West's *Guides*: 'Wordsworth's is close in substance to contemporary practical guidebooks of natural history [...] above all, Wordsworth is conscious of the shapes of the hills and the processes that continue to alter them' (*Wordsworth and the Geologists: A Correlation of Influences* (Cambridge: CUP, 1995), p. 24.

of Wordsworth's *Guide*, however, suggests that the old, static way of imagining landscape had not been abandoned; for all his imaginative wandering through the details of the Lucerne model, Wordsworth's chief interest in the diorama is that it supplies him with an elevated, fixed prospect which allows him to take in all the details at once.[47] It was not only foreign landscapes that Wordsworth treated in this way; his account of the basic organization of the Lake District uses the same techniques, including a static gaze from high above, to introduce the landscape that is to be the chief topic of the *Guide*.

Wordsworth announces that he will present a simplified graphic representation of the Lake country, which will be similar to the Lucerne model: it

will, in some instances, communicate to the traveller, who has already seen the objects, new information; and will assist in giving to his recollections a more orderly arrangement than his own opportunities of observing may have permitted him to make; while it will be still more useful to the future traveller, by directing his attention at once to distinctions in things which, without such previous aid, a length of time only could enable him to discover.[48]

The representation that Wordsworth is about to offer is intensely and explicitly spatial. I argued earlier that hub-and-ray images imagine the spatial arrangement as coming before, and as being more important than, the particular objects or information arranged, whereas aerial view images on the other hand tend to impose a spatial order on a pre-existing body of facts or objects. In this part of the *Guide*, Wordsworth confusedly attempts to use both kinds of image and hence to think of the spatial arrangement as pre-existing and at the same time post-dating the information it organizes. He hopes that his text will give the traveller who already knows the landscape a framework in which to arrange that knowledge (pre-existing objects being ordered by newly imposed spatial organization); equally, the traveller who has no information will be provided with an 'aid' which will help him to order his experience as he acquires it (spatial formation determining arrangement of newly-acquired objects). As a result of trying to hold together these two quite contrary methods of ordering, Wordsworth is obliged to create

---

[47] Michael Wiley notes that Wordsworth's *Guide* reproduces the picturesque vocabulary of 'stations' which Wordsworth inherited from writers including West: *Romantic Geography: Wordsworth and Anglo-European Spaces* (Houndmills: Macmillan, 1998), p. 152.

[48] Wordsworth, *A Guide through the Lakes*, pp. 170–1.

a very complicated temporal structure in this passage; his sentence is weighed down under a chain of tense shifts and indications of times: 'will communicate', 'has seen', 'will assist', 'have permitted', 'future traveller', 'at once', 'previous aid', 'length of time'.

Adopting a combination of the aerial view and the panoptical metaphor, Wordsworth sets about giving his reader the representation of the Lakes that he promised:

> To begin, then, with the main outlines of the country;—I know not how to give the reader a distinct image of these more readily, than by requesting him to place himself with me, in imagination, upon some given point; let it be the top of either of the mountains, Great Gavel, or Scawfell; or, rather, let us suppose our station to be a cloud hanging midway between those two mountains, at not more than half a mile's distance from the summit of each, and not many yards above their highest elevation; we shall then see stretched at our feet a number of vallies, not fewer than eight, diverging from the point, on which we are supposed to stand, like spokes from the nave of a wheel. First, we note, lying to the south-east, the vale of Langdale, which will conduct the eye to the long lake of Winandermere, stretched nearly to the sea; or rather to the sands of the vast bay of Morcamb, serving here for the rim of this imaginary wheel;—[...].[49]

Jonathan Bate comments:

> By substituting an imaginary station for an actual one, Wordsworth differentiates his *Guide* from those intended only for the bodies of tourists; with the image of the wheel, he introduces the idea of a unified place with a common centre. The remainder of Section First develops this sense of the unity of the country as formed by nature.[50]

The unity of the country is indeed, literally, formed by nature; but in another sense it is formed by being observed from the 'central perspective point' in the floating cloud. This is an example of a body of information having a structure imposed on it by the spatial image of a hub (in this case the observer's cloud) and rays (the valleys). But Wordsworth is concerned to emphasize the height required by the observer for the proper perception of the facts at his feet. To view the Lucerne model, the spectator 'ascends a little platform'; here, he is stationed hundreds of feet above the surface of the earth. The hub-and-ray structure as used

---

[49] Wordsworth, *A Guide through the Lakes*, p. 171. Michael Wiley convincingly reads the spokes in Wordsworth's imaginary wheel as owing an important debt to 'the sight lines taken by [William] Mudge as he made his trigonometrical measurements for the Ordnance Survey' (*Romantic Geography*, p. 158 and plate 8).

[50] Bate, *Romantic Ecology*, p. 46.

by Novalis or Coleridge is essentially two-dimensional. It would carry as much authority whether looked at from above, below or horizontally. Its strength is in its geometrical or abstract form. But Wordsworth's version of this image is three-dimensional, because it relies on adding height to the plane occupied by the hub and rays. By insisting on the three-dimensionality of his version of the image, Wordsworth compromises its abstraction and strengthens the power of the observer's position, rather than the power of the image itself.

There is playfulness in this passage, heard in the invitation to the reader to 'place himself with me' and again in 'let us suppose'. Wordsworth's spatial imagination here is not visionary, but fanciful. The winsomeness of the floating cloud image fits oddly with the literalism of his metaphor describing the vale of Coniston, which runs up 'from the sea, but not (as all the other vallies do) to the nave of the wheel, and therefore it may be not inaptly represented as a broken spoke sticking in the rim'.[51] Novalis and Coleridge use images of hubs and rays that lack an outer boundary or perimeter, whereas Wordsworth's metaphor has a 'rim' and retains an awareness of the practical utility of a wheel.[52] With this sense of the usefulness of a wheel goes the recognition of the potential for fracture or rupture: the vulnerable physicality of the object is preserved in Wordsworth's metaphor, whereas Coleridge's and Novalis's vastly powerful, universe-organizing hubs and rays are only barely physical and thus invulnerable.

The aerial view depends on establishing a distance between the observer and the world observed; this distance constitutes the observer's authority. Elizabeth Deeds Ermarth has argued that authority generated by a particular distance between the observer and the observed world is fundamental to realist representations: 'proper distance will enable the subjective spectator or the subjective consciousness to see the multiple viewpoints and so to find the form of the whole in what looks from a closer vantage point like a discontinuous array of specific cases'.[53] Aerial views rely on this 'proper distance'. But in the hub-and-ray model, the range of 'multiple viewpoints' is resolved into one single, ideal point by arranging the items into a geometric pattern,

51 Wordsworth, *A Guide Through the Lakes*, pp. 171–2.
52 For Wiley's reading of the 'rim' as metaphorically extended to take in the nation, see *Romantic Geography*, pp. 158–9.
53 Elizabeth Deeds Ermarth, *Realism and Consensus in the English Novel* (Princeton, N.J.: Princeton University Press, 1983), p. 35.

and distance or proximity becomes irrelevant, subsumed by symmetry and perspective. Both models finesse history, including the history of their own construction: both seek to order information so precisely and to place the reader so carefully in relation to that order that it becomes impossible to imagine the information organized differently. Wordsworth's aerial views erase the patterns of human life from the landscape; Coleridge's hubs and rays spell the final end of philosophical endeavour. Both these spatial organizations of knowledge aspire to conclusive, and concluding, authority.

Sorting and classifying, organizing data into an authoritative pattern, can be perilously close to snuffing out the vitality of the data, as Dorothea finds when her faith in Casaubon's Key to all Mythologies is shattered:

> And now she pictured to herself the days, and months, and years which she must spend in sorting what might be called shattered mummies, and fragments of a tradition which was itself a mosaic wrought from crushed ruins—sorting them as foison for a theory which was already withered in the birth like an elfin child.[54]

Not only the vitality of the data, but the vitality of the potential data-sorter herself, is in jeopardy. Because the principle on which the information has been selected is misguided, the information crumbles into dust, amid disturbing imagery of fertility gone awry. Mr Casaubon's desire was to show how 'all the mythical systems or erratic mythical fragments in the world were corruptions of a tradition originally revealed': in other words to arrange his apparently random material in such a way as to show its underlying unity and cohesion.[55] He believed in the Coleridgean doctrine that from a single privileged viewpoint all the random data of human knowledge would become comprehensible: 'Having once mastered the true position and taken a firm footing there, the vast field of mythical constructions became intelligible, nay, luminous with the reflected light of correspondences.'[56] Instead the pattern is discovered to be the projection of a fantasy of order, not order itself.

The dramatic irony at the end of chapter 48 of *Middlemarch*, where Dorothea is saved from having to take up Casaubon's lifework by finding him dead, is a corrective to Wordsworth's and Coleridge's fantasy of atemporal reading, of intuitive knowledge that demands utter

---

[54] Eliot, *Middlemarch*, p. 469 (Bk V, ch. 48).
[55] Ibid., p. 23 (Bk I, ch. 3).   [56] Ibid., p. 23 (Bk I, ch. 3).

conviction because of its Platonically perfect shape. Temporality always intervenes: it is only the passing of a few minutes that prevents Dorothea agreeing to Causaubon's demand to take on the burden of patterning his data. And where temporality intervenes, even spatial ordering can only ever be temporary. In this chapter Dorothea thinks that she has a privileged, overall view of her predicament ('she saw clearly enough the whole situation') but as soon as she begins to make her way towards Casaubon this privileged view disappears and she returns to the partial, surface-bound view of the ordinary human subject:

When she entered the Yew-tree Walk she could not see her husband; but the walk had bends, and she went, expecting to catch sight of his figure [...]. It occurred to her that he might be resting in the summer-house, towards which the path diverged a little. Turning the angle, she could see him [...].[57]

Middlemarch rejects the godlike 'central perspective point' of the atemporal visionary method of 'reading' proposed by Wordsworth and Coleridge, in their different ways. To Eliot the fantasy of such a perspective point is the result of egoism. Instead of Coleridge's and Novalis's great extending wheels that order all the things in the universe, Eliot perceives circles which point only to the self, and which are in any case an illusion:

Your pier-glass or extensive surface of polished steel made to be rubbed by a housemaid, will be minutely and multitudinously scratched in all directions; but place now against it a lighted candle as a centre of illumination, and lo! the scratches will seem to arrange themselves in a fine series of concentric circles round that little sun.[58]

It is tempting to read the contrast between Coleridge's and Eliot's epistemological circles as symptomatic of a Victorian disillusionment with the Romantic intellectual ambition that characterizes most of the writers discussed so far in this book. In the next chapter we shall see how spatial thinking helped early nineteenth-century writers to cut the field of human knowledge into smaller and more manageable pieces than would have been welcomed by either Coleridge, Wordsworth, or Casaubon.

[57] Ibid., pp. 472–3 (Bk. V, ch. 48).   [58] Ibid., p. 258 (Bk. III, ch. 27).

3

Disciplinary Boundaries and Border Disputes

Now the intellectual division of labour has always been an untidy affair—which is, in part, why there have been so many new disciplines and interdisciplinary projects brought into being—but it is not completely arbitrary. It is always possible to provide reasons (historical reasons) for the boundaries being drawn this way rather than that. Once those boundaries are established, however, they usually become institutionalized. All the apparatus of the academy is mobilized to mark and, on occasion, to police them. But these divisions do not correspond to any natural breaks in the intellectual landscape; social life does not respect them and ideas flow across them.[1]

In a sentence like 'my field of research is Tractarian poetry', the moribund metaphor 'field' sets up the sort of disciplinary boundaries on which modern academic life is founded. There are enormous numbers of spatial metaphors in use for this kind of purpose, and often they signify doubt rather than certainty about where to draw the lines between disciplines. For instance, the geographers Blakemore and Harley discuss 'the wide borderland between historical studies of art and cartography'.[2] Similarly, Arthur Eddington opens his *The Philosophy of Physical Science*: 'between physics and philosophy there lies a debatable territory which I shall call *scientific epistemology*'.[3] These metaphors

[1] Derek Gregory, *Geographical Imaginations* (Oxford: Basil Blackwell, 1994), pp. 10–11.

[2] M. J. Blakemore and J. B. Harley, 'Concepts in the History of Cartography: A Review and Perspective', *Cartographica* monographs 26, *Cartographica* 17.4, (Winter 1980), 1–120 (p. 76).

[3] Arthur Eddington, *The Philosophy of Physical Science* (Cambridge: CUP, 1939), p. 1.

divide up intellectual activity by reference to the kinds of spaces we understand in the physical world, and, like geographical borders, are routinely used to regulate who may enter particular spaces.

The presence of geographical metaphors in disciplinary discourse often indicates that the integrity of the discipline as a recognizable entity is at stake. In the early 1980s, for example, Edward Said described what he saw as the collapse of the integrity of the 'general field' of literary studies. He invoked a series of metaphors based on geography to bring to mind the division, invasion and distortion of disciplinary activity:

There seem to be too many interruptions, too many distractions, too many irregularities interfering with the homogeneous space supposedly holding scholars together. The division of intellectual labor, which has meant increasing specialization, further erodes any direct apprehension one might have of a whole field of literature and literary study; conversely, the invasion of literary discourse by the *outré* jargons of semiotics, post-structuralism, and Lacanian psychoanalysis has distended the literary critical universe almost beyond recognition.[4]

The idea of homogeneous space of *any* kind has come under attack from theorists of space, Foucault and Lefebvre chief among them.[5] It is certainly difficult to imagine what a 'homogeneous space' might be within the fissile and polyvocal world of modern literary studies. But if we change Said's disciplinary terms for scientific ones, replacing the 'whole field' of literature and literary study with a similar field of science, and semiotics *et al.* with branches of physics or chemistry, the passage closely resembles complaints very familiar from early nineteenth-century science writing.

Homogeneity played an important and affirmative role in early nineteenth-century science, including its sense of what we would now call disciplinarity. Interruptions, distractions, irregularities, and the division of intellectual labour were often seen as real threats to the attempt to develop a science that mirrored the homogeneity and uniformity of the natural world. And beyond the particular concerns of science, there

---

[4] Edward Said, 'Travelling Theory', in *The World, the Text, and the Critic* (Cambridge, Ma.: Harvard University Press, 1983; repr. London: Faber and Faber, 1984), pp. 226–47 (p. 228).
[5] Foucault suggests that the work of Gaston Bachelard decisively refutes the idea of homogeneous lived space ('Of other spaces', *Diacritics*, 16 (1986), 22–7, p. 23). Henri Lefebvre points out that even abstract space '*is not* homogeneous; it simply *has* homogeneity as its goal, its orientation, its "lens"' (*The Production of Space*, trans. Donald Nicholson-Smith (Oxford: Basil Blackwell, 1991), p. 287).

was the homogeneity of the general reading public to consider. The effect of mass access to education and specialized knowledge meant that the reading public could not be assumed to be a unified body or to be united in its basic assumptions. It is not surprising that resulting political questions about education and reading were often debated using strongly spatial discourses. Said's account of the state of literary studies is ironic, but his method of characterizing the discipline as a single, unified, nation-like space under attack by the triple threats of fragmentation, invasion, and distension is a legacy of early nineteenth-century writing about the organization of the intellectual world.

Disciplinary boundaries are always more or less under negotiation, but the first half of the nineteenth century was a time of particularly rapid and often controversial changes in disciplinarity. The traditional disciplines enshrined in university curricula were supplemented as the nineteenth century progressed by new courses in the sciences, modern languages, and social sciences. Further down the ladder of educational opportunity, the need to present knowledge to readers from a mass audience unfamiliar with traditional disciplinary distinctions stimulated critical attention to those distinctions.

In the first decades of the nineteenth century 'science' still carried its ancient meaning of knowledge or learning, so that definitions of 'science' often included disciplines which we would now consider to be in the humanities. For instance, Charles Babbage, the inventor of the difference engine and a great polemicist on scientific education, counted modern history, English law, and political economy as 'sciences' alongside chemistry, mineralogy, geology, and zoology.[6] Thomas Malthus saw 'science' as a spectrum in which his own area of expertise, 'the science of political economy', bore 'a nearer resemblance to the science of morals and politics than to that of mathematics'.[7] In some cases these broad definitions of science were supported by various kinds of political agenda. Growing optimism about the power and progressive effects of the experimental or inductive sciences meant that some writers sought to extend the beneficial methods of those sciences into the realm of polity, which involved shifting the disciplinary boundaries of 'science' to include a very wide variety of subjects. John Herschel, for

---

[6] Charles Babbage, *Reflections on the Decline of Science in England* (London: Fellowes and Booth, 1830; repr. Farnborough: Gregg, 1969), pp. 5–6.

[7] Thomas Malthus, *Principles of Political Economy* (1820), ed. John Pullen, 2 vols (Cambridge: CUP, 1989), I, p. 2.

example, argued that law and politics should be seen as 'experimental sciences' and history as 'the archive of experiments, successful and unsuccessful, gradually accumulating towards the solution of the grand problem—how the advantages of government are to be secured with the least possible inconvenience to the governed'.[8]

Both Babbage and Herschel were major figures in the debate which ran through the 1830s and 1840s over the disciplinary status of science as distinct from the arts (still usually meaning technologies or applied sciences) and the humanities, and about the statuses of the disciplines within science. As Herschel's radical attempt to apply scientific principles to the study of government suggests, this debate had strong social implications. The sciences had to negotiate their relationship with the developing public sphere. The emerging scientific disciplines (such as electromagnetics) needed to invent their own means of marking and policing their boundaries and at the same time to seek acceptance within the national and international cultures of science. The early history of the British Association for the Advancement of Science, founded in 1831, is a good demonstration of the contrary movements affecting the development of contemporary British science: a growing commitment to participation in public activities and public accessibility, and an increasing tendency for the disciplines within science to multiply and for regulatory power to be devolved to subject-specialists.

Modern accounts of the British scientific establishment in the early nineteenth century generally stress the increasing specialization of scientific practice during these years, emphasizing the growing division between disciplines and branches of science. David Knight, for instance, notes that by the fourth decade of the nineteenth century it was exceptional for a person devoting him/herself to science *not* to specialize in one particular branch.[9] The pace of change was sufficiently fast to provoke anxiety in some quarters that specialization might result in the total separation of one science from another: as we shall see, great concern was expressed that specialists would lose touch entirely with developments in other branches, and also, and perhaps more importantly, that the educated public would find it difficult or impossible to

---

[8] John F. W. Herschel, A *Preliminary Discourse on the Study of Natural Philosophy* (London: Longman, 1830), p. 73.

[9] David Knight, *The Age of Science: The Scientific World-View in the Nineteenth Century* (Oxford: Basil Blackwell, 1986), p. 4.

maintain an intelligent awareness of current opinion across the whole field of scientific research.

This apparent divergence of each science from the others was especially difficult to accept because disciplinary separation ran counter to the Romantic emphasis on the unity of nature and the growing stress on the universal reliability and applicability of the inductive method. Nature, for most commentators in this period, was 'uniform', an homogeneous field in which the same laws that operated within the sphere of human observation applied across the entire universe, so that the reasoning process which was valid for thinking about one part of nature was equally valid for all others.

In many cases this natural unity was taken to be the result of God's overarching and benevolent design of the universe, as for example by the astronomer and clergyman Temple Chevallier in 1836: 'it is no vague conjecture, but a highly probable conclusion upon strictly philosophical grounds [...] that one invisible, subtil, penetrating power, varying according to one simple law, pervades the whole material universe [...] and thus stamps the whole with the indelible impress of One great, powerful, intelligent, designing Mind'.[10] Similarly, William Hamilton Drummond's long poem *The Pleasures of Benevolence* (1835) includes a section in which an angel appears to the poet to explain that the uniformity of nature is a result of the goodness of God and that it is only man's partial view that makes some phenomena appear to be undesirable:

> Could'st thou survey
> Creation all—in whole and not in parts
> Discrete, disjointed, quickly would'st thou own
> That matchless wisdom planned,—that boundless good
> Pervades each movement of the fair machine[.][11]

And the American poet Philip Freneau makes the Popean point that

> No imperfection can be found
> In all that is, above, around,—
> All, nature made, in reason's sight
> Is order all, and *all is right*

[10] Temple Chevallier, *The Study of Mathematics as Conducive to the Development of the Intellectual Powers* (Durham: Parker, 1836), p. 26.
[11] William Hamilton Drummond, *The Pleasures of Benevolence: A Poem* (London: Wakeman, 1835), p. 109.

because of the uniformity of nature:

> On one fix'd point all nature moves,
> Nor deviates from the track she loves;
> Her system, drawn from reason's source,
> She scorns to change her wonted course.[12]

Both Freneau's and Drummond's poems use the image of nature as a machine rather than an organism, but a machine set in motion by God. The easy intermingling of religious and technoscientific discourses was a characteristic of this fascinating moment in cultural history, and one that made scientific disciplinary self-definition rather easier and less controversial than it might have been. The sharp antipathy between science and religion that existed, at least in popular perception, in the later nineteenth century, particularly following the publication in 1859 of *Origin of Species*, is not apparent to anything like the same extent or with the same degree of polarization in early nineteenth-century writing. As Jack Morrell and Arnold Thackray put it, in the early Victorian period, 'knowledge, religion, and learning formed an inseparable trinity in the eyes of educated men'.[13] This is not to say that there were not controversies over the effects that scientific and religious world-views had on their exponents and on readers. Morrell and Thackray show, for instance, that Newman, Keble, Froude, and other leaders of the Tractarian movement in the early 1830s opposed the British Association for the Advancement of Science because 'the BAAS was a new spirituality, a new claim to authority in affairs of the mind and of the State. That claim could only be met by displacing some other claimant, inevitably a religious one'.[14] Charles Babbage engaged in public controversy against William Whewell a few years over what Babbage attacked as an attempt to reclaim religion's ultimate authority on matters of nature. Babbage accused Whewell (who, as well as being ordained, was of course one of the great supporters of cutting-edge contemporary physical science) of reinforcing the prejudice that 'the

---

[12] Philip Morin Freneau, 'On the Uniformity and Perfection of Nature' (1815), in *Poems*, ed. Harry Hayden Clark (New York: Harcourt, Brace, 1929), p. 424, 423.

[13] Jack Morrell and Arnold Thackray, *Gentlemen of Science: Early Years of the British Association for the Advancement of Science* (Oxford: Clarendon Press, 1981), p. 224. See also pp. 225–9 for an excellent account of the liberal Anglican Christianity at the centre of the developing BAAS.

[14] Ibid., p. 231.

pursuits of science are unfavourable to religion'.[15] And throughout the period, some kinds of pro-religious polemic continued to oppose even the most generally accepted scientific knowledge. While major national figures debated the limits of scientific and religious claims to authority, small publishers were producing a stream of treatises arguing against a scientific account of the universe in favour of a Biblical one. John Finleyson's 1830 work, *The Universe As It Is, and the Detection and Refutation of Sir Isaac Newton; Also, the Exposure and Proved Fabrication of the Solar System*, was extreme but not egregious in using a wealth of scriptural quotations to urge a return to a pre-Copernican belief in the rotation of the sun around the earth.[16] But in the rhetoric of most mainstream British writers on science, orthodox if sometimes unzealous religious faith ran alongside up-to-date scientific information and outlooks. In the work of influential commentators such as William Whewell and Mary Somerville, and popularizers like Henry Brougham, belief in the urgency of scientific progress was combined with explicit reverence for God, whose creation nature was and whose omnipresence and oneness underlay the homogeneity of natural laws.

ASTRONOMY, HOMOGENEITY, AND VISION

The scientific reverence which connected God with the great forces of the natural world was partly fuelled by the discourse of the sublime, which allowed the emotional impact of awe and wonder to slide easily between allegiances to religion and to science. The connection of religion and science via the sublime is clear in contemporary writing about astronomy, for instance. The immensity and beauty of the night sky meant that astronomy was considered particularly likely to expand the mind's capacity for wonder and to elevate the character. The

---

[15] Charles Babbage, *The Ninth Bridgewater Treatise: A Fragment* (London: John Murray, 1837), p. x.

[16] John Finleyson, *The Universe As It Is, and the Detection and Refutation of Sir Isaac Newton; Also, the Exposure and Proved Fabrication of the Solar System* (London: Snell, 1830). B. Prescot's slightly earlier but similar argument was contained in a book whose title told its whole story: *The Inverted Scheme of Copernicus; with the Pretended Experiments upon which his Followers have Founded their Hypotheses of Matter and Motion, Compared with Facts, and with the Experience of the Senses: and the Doctrine of the Formation of Worlds out of Atoms, by the Power of Gravity, Contrasted with the Formation of One World by Divine Power, as it is Revealed in the History of the Creation* (Liverpool and London: Riebau, 1822).

contemplation of outer space was often argued to be the most uplifting of all scientific subjects: as Mary Somerville put it, 'the heavens afford the most sublime object of study which can be derived from science'.[17] As her term 'the heavens' indicates, vestiges of a pre-scientific association of the sky with the divine lingered in the vocabulary of discursive astronomy, making it easy to present the science as a pious one.

Of the illustrations given by nineteenth-century scientific writers of the homogeneity of the universe, astronomy was generally agreed to be one of the most impressive. It linked the individual human mind with the most remote and immense events observable in the cosmos, showing that science was capable not only of understanding but even of foretelling these events. The universe must really be an homogeneous field if human reason had such triumphantly wide applicability. Astronomy was also often pointed up as an example of a superbly successful scientific discipline, and the connection between success as a discipline and a tendency to highlight the unity of nature was not coincidental. Astronomy broadened and dignified the mind of the astronomer, combined observation and abstraction in satisfying proportions, had a claim on everyone's aesthetic sensibilities via the beauty of the stars, and made sense not only of past observations but had a grip on the future as well. James Nicol, Professor of Natural History at the University of Aberdeen, wrote in his *Introductory Book of the Sciences* in 1844 that 'no more ennobling or exalted object can be presented to the contemplation of the human mind than the science of Astronomy, which, by means of a few simple laws, explains all the phenomena of the heavens, enables us to foresee them long before they occur, and points out their connexion with facts that are every day observed on the earth.'[18]

Nicol alludes to the factor which made astronomy a favourite example of a successful scientific discipline. Not only was astronomy readily compatible with piety, but it was also the science which had yielded the most dramatic example of the power of scientific reasoning: its power of prediction. Emphasizing that the future was now accessible via science had the subtle and, in many cases, unconscious, effect of giving science some of the authority that religion had claimed.

---

[17] Mary Somerville, *On the Connexion of the Physical Sciences* (London: John Murray, 1834), p. 3.
[18] James Nicol, *Introductory Book of the Sciences, Adapted for the Use of Schools and Private Students* (Edinburgh: Oliver and Boyd; and London: Simpkin, Marshall, 1844), p. 143.

So impressive was science's power to predict future astronomical events that a religious or at least mystical discourse was sometimes thought appropriate for it. For instance, in crowd-pleasing rhetoric in an address to the third meeting of the British Association in 1833, William Whewell argued that astronomy is 'the queen of sciences' and the only 'perfect' one, 'the only branch of human knowledge in which particulars are completely subjugated to generals, effects to causes;—in which the long observation of the past has been, by human reason, twined into a chain which binds in its links the remotest events of the future;—in which we are able fully and clearly to interpret nature's oracles, so that by that which we have tried we receive a prophecy of that which is untried.'[19] Whewell's references to oracles and prophecies appropriate the predictive authority of religion, though, characteristically, Whewell discreetly suggests that it is pagan rather than Christian religion which has been overtaken by science. Astronomy is used as an example of science's ability to redeem the past, to turn its own history to good account, as the much more liberal Herschel hoped that eventually all history would be turned, so that instead of being 'the mere record of tyrannies and slaughters' it would be recognized as 'the archive of experiments [...] towards the solution of the grand problem' of just government.[20]

John Herschel, whose own reputation in astronomy complemented that of his aunt and his father, the discover of Uranus, recognized in his 1830 *Preliminary Discourse* that even in pre-scientific times, star-watchers had been lent immense authority by their ability to predict eclipses: this predictive power was 'one grand instrument by which [mankind's] allegiance (so to speak) to natural science, and their respect for its professors, has been maintained'.[21] By means of prediction, the impressiveness of the behaviour of the stars could be presented as a spectacular demonstration of the authority of knowledgeable humans. Accurate prediction of the movements of the stars and the return of comets, for Herschel, was a powerful tool in the public rhetoric of science even in his own age in which 'men prefer being guided and enlightened, to being astonished and dazzled'.[22]

---

[19] William Whewell, *Address Delivered in the Senate-House at Cambridge, June XXV, MDCCCXXXIII. On the occasion of the opening of the third General Meeting of the British Association for the Advancement of Science* (Cambridge: Smith, 1833), pp. 5–6.
[20] Herschel, *Preliminary Discourse*, p. 73.
[21] Ibid., p. 27.   [22] Ibid., p. 27.

A decade and a half after writing these sentences, Herschel was one of the astronomers involved in the discovery of Neptune, which was first observed in 1846 but which had been predicted since at least 1841. The prediction was based on the fact that observations of perturbations in the orbit of Uranus—then thought to be the most distant planet in the solar system—could not be accounted for by reference to the gravitational effects of the other known planets. Another body, further out than Uranus, must be acting on it. In September 1846, Herschel told the British Association for the Advancement of Science of this surmised body that 'we see it as Columbus saw America from the shores of Spain. Its movements have been felt, trembling along the far-reaching line of our analysis, with a certainty hardly inferior to that of ocular demonstration.'[23] Two weeks later, the German astronomer Galle visually observed the new planet in the location where mathematical calculations had predicted it must be. Almost immediately, this triumphal feat of mathematical astronomy became a symbol of the successful application of scientific reasoning beyond the domain of the senses. As Herschel wrote in the *Edinburgh Review* in 1848,

The discovery of [...] Neptune, by the mere consideration of the recorded perturbations of the remotest planet previously known, by the theory of gravitation, as delivered by Newton, and matured by the French geometers, will ever be regarded as the most glorious intellectual triumph of the present age.[24]

And in 1855 Michael Faraday argued that the discovery of Neptune—'an orb rolling in the depths of space, so large as to equal nearly sixty earths, yet so far away as to be invisible to the unassisted eye'—showed that astronomy's gravitation-based knowledge was second only to the word of God in certainty, asking: 'what truth, beneath that of revelation, can have an assurance stronger than this?'[25]

The predictive power of astronomy powerfully suggested that the homogeneity of the universe was temporal as well as spatial: that is, that there were no sudden overturnings of natural laws. What was true

---

[23] J. F. W. Herschel, 'Presidential Address', *British Association for the Advancement of Science Reports* (1846), pp. xxviii–xliv.

[24] [J. F. W. Herschel], review of Humboldt's *Kosmos*, *Edinburgh Review* 87 (January 1848), 170–229 (191).

[25] Michael Faraday, 'On Mental Education', in *Experimental Researches in Chemistry and Physics* (London: Taylor & Francis, 1859), pp. 463–91, p. 470. Faraday goes on to contrast the certainty of the law of gravity with its disregard among those who believe in spiritualist manifestations.

today would be equally true tomorrow, just as what was true here would also be true out there. If this were not the case, no prediction based on past observations could be relied on. The mid-Victorian classicist, mathematician, and unitarian Francis Newman, brother of John Henry, made this point explicit in a poem of 1858:

> Experience of the past avails not for knowing the future,
> Unless Uniformity of Natural Law be *presumed*,
> Which has no direct demonstration, and is at first matter of Faith.
> Yet this *Faith gives foresight* to the untaught and unphilosophic,
> And the reasoner, who disowns it, is but foolish or insane.[26]

Newman's point was that even unscientific peoples were able to make reasonably accurate predictions about tides and astronomical events because they had an instinctive belief in the uniformity of nature. His larger argument extended this idea of the reliability of natural laws to the reliability of moral ones, which lead us to believe that 'Evil must wear itself out, and Good bear final sway, | That the Future is big with blessing, which the triumph of Good shall unfold'. To believe in the homogeneity of nature must be to believe in the goodness of 'God's laws', and the predictive power of astronomy is a beneficial result of both.[27]

Nature's uniformity was not the result of human perception, and was, for some commentators, quite independent of observation. As Mary Somerville wrote, humanity could and would be entirely destroyed but natural law would remain:

man is also to vanish in the ever-changing course of events. The earth is to be burnt up, and the elements are to melt with fervent heat—to be again reduced to chaos—possibly to be renovated and adorned for other races of beings. These stupendous changes may be but cycles in those great laws of the universe, where all is variable but the laws themselves [...].[28]

Somerville's confident account of the future prospects of the earth combines narrative elements which are compatible with Biblical predictions of the Last Judgement with others that are not—notably, the suggestion that a remade earth might be inherited by 'other races of beings'. The assumption of temporal homogeneity threatened to bring astronomy's

---

[26] Francis William Newman, 'Faith and Foresight', in *Theism, Doctrinal and Practical; or, Didactic Religious Utterances* (London, 1858), p. 57.
[27] Ibid., p. 58.
[28] Mary Somerville, *Physical Geography*, 2 vols (London: John Murray, 1848), I, p. 3.

claims—and the claims of those sciences that used astronomy as a model of success—alarmingly close to conflict with Biblical narrative, and as the earth sciences were discovering at this period, any such conflict had to be handled with great care.[29] But it is characteristic not only of Somerville's writing but of much early nineteenth-century comment on the authority of science that the ultimate link in the chain that binds humans and the universe together should turn out to be God. Somerville concludes her sentence: 'all is variable but the laws themselves and He who has ordained them'.[30] God is the guarantor of the homogeneity of the field of nature. Because God is omnipresent and unchanging, so are his laws.

The reliable homogeneity of nature was clearly in many ways a deeply consoling idea, but one compatible with progressive as well as conservative ideologies, both political and religious. Assuming nature's uniformity made extraordinary leaps of the scientific imagination possible, because even parts of nature that were inaccessible to observation via the senses could be explored via reasoning. This in turn had an effect on early nineteenth-century concepts of disciplinarity, setting up a fundamental distinction between those sciences that could not be done without observation (such as biology) and those that relied profoundly on reasoning (such as geometry).

Whether or not God was explicitly invoked as the underlying cause for the homogeneity of the field of nature, it was the abstract reasoning which explored that homogeneity that gave astronomy its predictive power. Mere massing up of observations could not achieve this power: local instances had to be converted into universal law before prediction could be accurate, and this meant shifting away from the evidence of the senses to that of the inductive reasoning process. John Herschel's claim that the certainty of mathematical prediction was 'hardly inferior' to the certainty of the senses was an understatement of the confidence of nineteenth-century scientific reasoning. Supporting this confidence was the basic faith that mathematics as it had come to development in Europe profoundly matched the structure of the universe, so that mathematical operations, quite independent of sense data, could accurately reflect the

---

[29] For example, the controversy over vulcanism had significant effects on the relationship between geology and religion in this period. For discussion see Charles Coulston Gillispie, *Genesis and Geology: A Study in the Relations of Scientific Thought, Natural Theology and Social Opinion in Great Britain, 1790–1850* (London: Oxford University Press, 1951).

[30] Somerville, *Physical Geography*, I, p. 3.

workings of the universe. Mathematics could express the laws of the physical world, and could perhaps go where the human senses might never catch up.

Mathematics had always had a very high status among the academic subjects, but in the early nineteenth century this status was complicated and questioned by changing patterns of education resulting from the expansion of literacy and the promotion of education for the artisan class. Along with classics, mathematics was at the core of a gentleman's education, and it dominated the Cambridge University curriculum which produced Babbage, Herschel, Whewell, and many of the other key figures in the British scientific establishment in the early nineteenth century. As one Victorian historian of mathematics noted, 'under the influence of the Newtonian philosophy mathematics gradually became the dominant study of the place [Cambridge], and for the latter half of this time the mathematicians controlled the studies of the university almost as absolutely as the logicians had controlled those of the medieaeval university'.[31] The result was that although other subjects could be studied, it was not until mid-century that Cambridge students could take another BA Honours degree without first taking the BA exam in mathematics.[32] But because it was unusual for women and non-public school educated men to be taught more than arithmetic, the study of mathematics was a marker of class and gender boundaries which served to exclude many from study of the sciences such as astronomy which depended on mathematical operations.[33] From the late 1820s, though, the pressure of changes in the educational needs and aspirations of men and boys of the middle and artisan classes led to the publication of a large number of textbooks on mathematics and mathematical sciences aimed at this new market, though it was still assumed that these were almost exclusively masculine requirements.

Mathematics was often presented as the most universal of the sciences, since it did not depend on experience or observation but on intellectual

---

[31] W. W. Rouse Ball, *A History of the Study of Mathematics at Cambridge* (Cambridge: CUP, 1889), pp. 253–4. Newton had of course been Lucasian Professor of Mathematics at Cambridge from 1669–1701.

[32] Rouse Ball, *A History of the Study of Mathematics at Cambridge*, p. 212.

[33] An obvious early nineteenth-century exception is Ada Lovelace, whose mother Annabella, Lady Byron shared her own knowledge and enthusiasm for mathematics with her daughter. For an account of Annabella's and Ada's mathematical education see Judith S. Lewis, 'Princess of Parallelograms and Her Daughter: Math and Gender in the Early Nineteenth Century Aristocracy', *Women's Studies International Forum*, 18 (1995), 387–94 (pp. 389–92).

powers only. The severe barriers surrounding access to the study of mathematics contributed to the cultural mystique that it acquired and, paradoxically, to its claim to universalism. The success of the mathematical operations that allowed astronomers to predict the position of an as yet unobserved planet were an example of mathematics's independence of visual and all other sense information. Astronomy was thus often considered to exemplify the ideally rational rather than sensory basis of the physical sciences, and to be the most complete application of mathematical techniques to the study of the natural world.

This dependence on mathematics could be, and was, used as a means of bracketing some kinds of science off from the others and thus forming a basis for a categorization of the scientific disciplines. The homogeneous field of the natural world presented itself in different ways to the human mind, and the division between those sciences that did and did not require observation of phenomena was for many early nineteenth-century writers on science the most fundamental one. It was the basis of the disciplinary distinctions made by a wide range of well-known and influential writers. John Herschel's category of 'abstract science', for example, relied on exactly this difference:

Abstract science is independent of a system of nature,—of a creation,—of every thing, in short, except memory, thought, and reason. Its objects are, first, those primary existences and relations which we cannot even conceive not to *be*, such as space, time, number, order, &c.; and, secondly, those artificial forms, or symbols, which thought has the power of creating for itself at pleasure, and substituting as representatives, by the aid of memory, for combinations of those primary objects and of its own conceptions [...].[34]

With only memory, thought and reason to rely on, abstract science could have little use for the senses. Romantic science tended to follow German idealism in emphasizing the value of the rational over the empirical. Even Alexander von Humboldt, whose immense contribution to nineteenth-century science had been based very largely on his observations of plants and geological formations all over the world, wrote that

the work of the intellect shews itself in its most exalted grandeur, where, instead of requiring the aid of outward material means, it receives its light exclusively from the pure abstraction of the mathematical development of thought. There dwells a powerful charm, deeply felt and acknowledged in all antiquity, in

---

[34] Herschel, *Preliminary Discourse*, p.18.

the contemplation of mathematical truths; in the eternal relations of time and space, as they disclose themselves in harmonics, numbers, and lines.[35]

Because mathematics distinguished itself from the other sciences by relying on rational rather than empirical processes, and had objects which could be thought of as ideal or imaginary rather than visual or tangible, it could be studied without access to observations of the external world. Sciences such as botany, biology or chemistry could not. Herschel made this point explicitly in his *Preliminary Discourse*:

A clever man, shut up alone and allowed unlimited time, might reason out for himself all the truths of mathematics, by proceeding from those simple notions of space and number of which he cannot divest himself without ceasing to think. But he could never tell, by any effort of reasoning, what would become of a lump of sugar if immersed in water, or what impression would be produced on his eye by mixing the colours yellow and blue.[36]

De Quincey used the same premise to categorize what he called the objective and subjective sciences. The objective sciences were those which carry the mind 'to materials existing *out* of itself, such as natural philosophy, chemistry, physiology, astronomy, geology'. The subjective ones were those which could be studied by a man isolated from natural phenomena and from other human beings: 'rather than quit the external world, he must in his own defence, were it only as a relief from gnawing thoughts, cultivate some *subjective* science; that is, some branch of knowledge which, drawing everything from the mind itself, is independent of external resources.' De Quincey's examples of this category were theology, metaphysics, logic, and geometry, and although he admitted that sometimes the devotees of the subjective sciences could waste time erecting 'Jacobs's ladders of vapoury metaphysics', nonetheless he argued that it was only from these studies that 'the culture of style' could emerge.[37]

Henry Brougham made the same distinction between mathematics and sciences that require observation in his very widely reproduced pamphlet 'A Discourse of the Objects, Advantages and Pleasures of

[35] Humboldt, *Cosmos: Sketch of a Physical Description of the Universe*, transl. under the superintendence of Edward Sabine, new edn, 5 vols (London: Longman and Murray, 1856), II, pp. 351–2.

[36] Herschel, *Preliminary Discourse*, p. 76.

[37] De Quincey, 'Style. No. IV.' (1841), in *The Works of Thomas De Quincey*, gen. ed. Grevel Lindop, 21 vols (London: Pickering and Chatto, 2000–3), 12 (2001), pp. 64–84 (pp. 66–7, 71).

Science', the first publication of the newly-formed Society for the Diffusion of Useful Knowledge in 1827. In a passage which seems to have been a direct influence on Herschel's discussion of the disciplines in his *Preliminary Discourse*, Brougham argued that one of the basic boundaries between disciplines was that which separated studies that demanded experimentation and observation from those which could be done in conditions of complete isolation from the physical world:

> If a man were shut up in a room with pen, ink, and paper, he might by thought discover any of the truths in arithmetic, algebra, or geometry; it is possible, at least; there would be nothing absolutely impossible in his discovering all that is now known of these sciences; and if his memory were as good as we are supposing his judgment and conception to be, he might discover it all without pen, ink, and paper, and in a dark room. But we cannot discover a single one of the fundamental properties of matter without observing what goes on around us, and trying experiments upon the nature and motion of bodies.[38]

Because of his public position and because of the politics of the SDUK's intervention in publishing for the mass reading public, Brougham's 'Discourse' attracted an enormous amount of attention, not all of it favourable either to its politics or its science. In a substantial and pointed satire called *The Blunders of a Bigwig; or, Paul Pry's Peeps into the Sixpenny Sciences* published almost immediately after the appearance of the 'Discourse', 'Paul Pry' took particular issue with Brougham's distinction between sciences requiring visual observation and those that do not:

> Mr B. thinks it impossible for a person who is blind, to discern the Laws of Nature by other senses than sight. But, if he had been less precipitous and dogmatical, he might probably have recollected, that Professor Saunderson gave lectures even in Optics, and the Theory of Colours, with infinite applause before the University of Cambridge, and made many valuable discoveries.[39]

The Lucasian Professor of mathematics at Cambridge from 1711 until his death from scurvy in 1739, Nicholas Saunderson had been blind since a childhood attack of smallpox. Dealing with subject matter which his students experienced visually but which to him could only be a topic of rational investigation, Saunderson's achievements in empirical

---

[38] [Henry Brougham, 1st Baron Brougham and Vaux], *A Discourse of the Objects, Advantages, and Pleasures of Science*, 2nd edn (London: Baldwin, Cradock, and Joy, 1827), pp. 14–15.
[39] 'Paul Pry', *The Blunders of a Bigwig [...]* (London: Hearne, 1827), p. 11.

science were extraordinary. Half a century after Saunderson's death, Edmund Burke commented:

This learned man had acquired great knowledge in natural philosophy, in astronomy, and whatever sciences depend upon mathematical skill. What was the most extraordinary [...], he gave excellent lectures upon light and colours; and this man taught others the theory of these ideas which they had, and which he himself undoubtedly had not. But it is probable that the words red, blue, green, answered to him as well as the ideas of the colours themselves; for the ideas of greater or lesser degrees of refrangibility being applied to these words, and the blind man being instructed in what other respects they were found to agree or to disagree, it was as easy for him to reason upon the words, as if he had been fully master of the ideas.[40]

Burke's account of Saunderson's method depends once again on making a distinction between mathematics and the other sciences. Because astronomy, despite its reliance on visual observation, depended very heavily on mathematics, it could be practised by someone who could not see the stars but who could undertake mathematical operations in the sensorium of his mind.

It was probably for this reason that Robert and William Chambers chose astronomy as the subject of one of their short-lived 1840s series of books for the blind. Printed in embossed capital letters (the twelve-year-old system of Braille was not yet in common use), the text was simply that of the Chambers brothers's *Introduction to the Sciences, for the Use of Schools and for Private Instruction* and was totally unadapted for its blind readership. The Chambers did not scruple to reproduce passages that were quite inappropriate for the blind: for instance, 'to make it clear that the earth is round, we may, on a clear day, look out from some high ground upon the sea, when we shall see the tops of approaching vessels first appear, & gradually the lower parts'.[41] The only other volume in the series was a guide to the London sights, again reprinted from a work for those without visual impairment, a choice which suggests that the intention of the series was to compensate for

---

[40] Edmund Burke, *A Philosophical Inquiry into Our Ideas of the Sublime and Beautiful* (1757), ed. David Womersley (London: Penguin, 1998), p. 192 (Part V, ch. 5).

[41] W. and R. Chambers, *Introduction to the Science of Astronomy, Embossed by Permission, for the Use of the Blind* (Glasgow: John Smith *et al.*, 1841), p. 3. *Introduction to the Sciences: For Use in Schools and for Private Instruction* (Edinburgh: Chambers, 1838).

blindness by verbal description.[42] Nonetheless, the astronomy volume's appeals to visual experience show a strange insensitivity to its readers.

As these examples indicate, early nineteenth-century disciplinary divisions, with all their consequences for authority claims, were caught up in much wider discussions of the role of sensory experience and abstract rationality in science. Systems like Herschel's and Brougham's located fundamental disciplinary coherence in distinctions based on the necessity or otherwise of empirical data. They were variations on a disciplinary hierarchy which saw as the most fundamental sciences those that depended least on empirical data. Those such as chemistry, for instance, which depended heavily on experimental data rather than mathematical reasoning were prone to devalorization as a result.[43]

DISCIPLINARY BOUNDARIES AND THE STATE OF SCIENCE

The characteristic effort of early nineteenth-century science was to understand natural phenomena by ever more general laws. An account of the universe as operating by a smaller rather than a larger number of laws seemed neater, more elegant, and more aesthetically pleasing, and hence more likely to reflect accurately the values really embodied in nature. Charles Babbage was voicing a very widely held view when he wrote in 1837 that

All analogy leads us to infer, and new discoveries continually direct our expectation to the idea, that the most extensive laws to which we have hitherto attained, converge to some few and general principles, by which the whole of the material universe is sustained, and from which its infinitely varied phenomena emerge as the necessary consequences.[44]

---

[42] W. and R. Chambers, *A Description and Guide to the British Metropolis; Embossed by Permission for the Use of the Blind* (Glasgow: John Smith *et al.*, 1841).

[43] Mary Jo Nye sums up contemporary chemistry's situation in the hierarchy of disciplines: 'nineteenth-century scientists and philosophers grew accustomed to claiming that chemistry is a science of taxonomy and classification resting on empirical and descriptive foundations, in contrast to mechanical physics resting on axiomatic and mathematical foundations' (*From Chemical Philosophy to Theoretical Chemistry: Dynamics of Matter and Dynamics of Disciplines, 1800–1950* (Berkeley: University of California Press, 1993), p. 5).

[44] Babbage, *Ninth Bridgewater Treatise*, p. 32.

Many writers on science extended this belief to argue that in the distant future the ultimate goal might be reached of expressing a single sublime law comprehending all the workings of the universe. Mary Somerville had perhaps her greatest influence with *On the Connexion of the Physical Sciences*, a book attempting to demonstrate how phenomena that seemed diverse and unrelated were now or would be in future explicable by the same natural laws. Somerville's book contributed to the currency of the idea that even the apparently basic and separate forces of light, heat, electricity, and magnetism would 'ultimately be referred to some one power of a higher order, in conformity with the general economy of the system of the world, where the most varied and complicated effects are produced by a small number of universal laws'.[45] Far from suggesting impiety, it was generally assumed that striving towards this law would bring humanity closer to the divine mind.

Even pseudo-science often accepted this ideology of natural unity. T. S. Mackintosh, whose *'Electrical Theory'* argued that the human capacity for being attracted to pleasing things and repulsed from painful ones was a result of magnetism, demonstrating a fundamental misunderstanding of contemporary scientific knowledge of electromagnetic force, nonetheless was clearly swayed by the general trend of scientific ambition when he announced that his project was

to endeavour to discover and determine ONE UNIVERSAL LAW, that shall comprehend every action and re-action, both physical and moral. In short, that shall embrace every variety of phenomena, in matter and mind, which is presented to our notice in the wide field of nature.[46]

But this goal of a single law only made sense if the universe could be assumed to be homogeneous, with no pockets of difference and no random anomalies, so that all parts of the universe were subject to the same physical laws and could be described by the same kind of science. The tendency of Romantic science, as we have seen, was to assume this homogeneity; it was this assumption that provided the science of this period with much of its optimism. Phenomena that seemed wholly unconnected might ultimately be found to be simply various workings of a single, deep force, and this suggested that the spaces in which these phenomena existed were all part of a single continuum.

---

[45] Somerville, *On the Connexion*, p. 358.
[46] T. S. Mackintosh, *The 'Electrical Theory' of the Universe. Or, The Elements of Physical and Moral Philosophy* (London: Simpkin, Marshall; and Manchester, Heywood [*c*.1838]), p. 7.

The earth was not a uniquely privileged part of the universe but one world among myriads, and the laws which governed this planet applied equally to other bodies far out in space. Similarly, the individual human and the human species, being themselves phenomena, were an intimate though tiny element of the universe and participated in its patterns. As the Danish physicist and Romantic scientific essayist Hans Christian Oersted put it, the scientifically-educated person could feel 'that he is accepted as a participating link in the eternal principle of the universe'.[47]

The belief in an homogeneous universe, in which everything could be explained by a small number of fundamental laws, suggested that the contemporary fragmentation of scientific disciplines was only a temporary stage in the onward march of science. Although a great deal of rhetoric and much intellectual effort was expended on promoting the unification of the disciplines, most scientific practitioners recognized that specialization was the way to increase the knowledge of phenomena which might well one day be seen as different facets of a single principle but which must currently be studied—at least to some extent—in isolation from each other. There remained considerable difficulty as to how to represent this situation to the educated public, who wanted to keep abreast of contemporary technological progress and who ought to benefit from the uplifting moral character of sublime scientific knowledge. As I discussed in the last chapter, projects such as encyclopaedias had among their aims that of encapsulating fast-diverging branches of knowledge, structuring them in such a way that the reader could make sense of their relation to one another. But the first decades of the nineteenth century also saw the growth of scientific popularization in many other forms.

SOMERVILLE, WHEWELL, AND THE DIVIDED DISCIPLINARY LANDSCAPE

Mary Somerville's *On the Connexion of the Physical Sciences* was a text much concerned with boundaries, disciplinary demarcation, and the understanding of space both terrestrial and cosmological. Although professedly intended for women—that is, interested amateurs rather

---

[47] Hans Christian Oersted, 'The Comprehension of Nature by Thought and Imagination', in *The Soul in Nature: With Supplementary Contributions*, trans. Leonora and Joanna B. Horner (London: Bohn, 1852), pp. 41–55 (p. 55).

than specialist professionals—the book received much praise for its ability to pull together scientific research from a number of different branches in such a way as to be of professional interest to scientific specialists. Already known and highly respected in scientific circles for her 1831 translation and explanation of Laplace's *Traité de mécanique céleste*, Somerville's reputation was greatly enhanced by this survey of the current state of knowledge in a variety of scientific disciplines. *On the Connexion* diligently presents itself as not attempting to cross the heavily policed boundary between the sexes; though written by a woman, a fact which of course received a good deal of attention from reviewers, nonetheless it was careful not to claim authority over a male audience. Dedicating the book to Queen Adelaide, Somerville sought 'to make the laws by which the material world is governed more familiar to my countrywomen'.[48] According to Thackeray in *Vanity Fair*, a section of Somerville's countrywomen responded warmly enough to make her the fashion: 'some of the ladies were very blue and well informed; reading Mrs. Somerville, and frequenting the Royal Institution'.[49] Reviews in the non-scientific periodical press, which were generally very favourable, frequently took Somerville's claim that her book was intended for women at face value and suggested that men should be encouraged to read it too.[50]

Beyond its cautious negotiation of gender, *On the Connexion of the Physical Sciences* is an interesting starting point for a consideration of pre-Victorian scientific boundaries. It participates in the urgent contemporary debate about the progress of science into disciplinarity and whether increasing specialization or a renewed aspiration to inclusiveness should dominate the practice and profession of science. At the same time as Somerville stresses the limitations of her own role as popularizer and as woman writer in the field, her argument strongly supports a belief in a science free of constricting and artificial limits. Her preface explains the underlying assumption of the book as promoting a conception of an organic, holistic science without internal boundaries:

[48] Somerville, *On the Connexion*, n. p., dedication.
[49] William Makepeace Thackeray, *Vanity Fair* (1848), ed. John Carey (London: Penguin, 2001), p. 720 (ch. 61).
[50] I have discussed this book and its place in Somerville's career elsewhere: see Alice Jenkins, 'Writing the Self and Writing Science: Mary Somerville as Autobiographer', in Juliet John and Alice Jenkins, eds, *Rethinking Victorian Culture* (Houndmills: Macmillan, 2000), pp. 163–79.

The progress of modern science, especially within the last five years, has been remarkable for a tendency to simplify the laws of nature, and to unite detached branches by general principles. In some cases identity has been proved where there appeared to be nothing in common, as in the electric and magnetic influences; in others, as that of light and heat, such analogies have been pointed out as to justify the expectation, that they will ultimately be referred to the same agent: and in all there exists such a bond of union, that proficiency cannot be attained in any one without a knowledge of others.[51]

Most of the reviews accepted Somerville's account of the state of science. As part of her express (though perhaps somewhat disingenuous) rhetoric, Somerville described her own project in modest terms, which made it easier for reviewers to concur with her fairly radical view of the nature of science. The *Edinburgh Review*, for instance, which was lavish in its praise of the volume, argued that

it does not form any part of her plan to establish the identity of particular branches of knowledge, or to trace any bond of union by which they may be mutually enchained; or to point out the means by which the cultivation of any one science may lead to the extension of another. Such a discussion, however interesting in itself, and important to the progress of science, would have led her beyond the sphere of a popular treatise, and would have frustrated the principal object which she had in view.[52]

In other words, the *Edinburgh*'s reviewer responded warmly to the limits Somerville placed on her aims; interestingly, he did so by turning her own vocabulary back on her, adopting but re-deploying her phrase 'bond of union' as just one of a string of metaphors dominating his disciplinary rhetoric.

The most important discussion of Somerville's book, though, appeared in the *Quarterly Review* and was by the classical philologist, mineralogist, Fellow of Trinity College, Cambridge and polymath William Whewell.[53] The *Quarterly*'s editor, John Lockhart, offered the project of reviewing Somerville's book to Whewell in what Richard Yeo rightly calls 'terms that could hardly be described as eloquent', indicating that he had thought of John Herschel and David Brewster as

---

[51] Somerville, *On the Connexion*, n. p., preface. See also pp. 358–9 for further elaboration of the ultimate unity of the sciences.

[52] *Edinburgh Review*, 59 (April–July 1834), 154–71 (p. 155).

[53] [William Whewell], unsigned review, *Quarterly Review*, 51 (March 1834), 54–68.

possible reviewers but was obliged to settle for Whewell.[54] Despite this unpromising start, the resulting review was one of a number of agenda-setting pieces Whewell wrote for the periodical press in the early part of the 1830s, pieces that Yeo describes as 'metascientific commentaries' in which Whewell developed his views on 'the meaning of "science" both within the scientific community and for the public'.[55]

Like many of Somerville's reviewers, Whewell took the opportunity to comment on the suitability of women to engage in science:

Notwithstanding all the dreams of theorists, there is a sex in minds. One of the characteristics of the female intellect is a clearness of perception, as far as it goes: with them, action is the result of feeling; thought, of seeing; their practica[l] emotions do not wait for instruction from speculation; their reasoning is undisturbed by the prospect of its practical consequences. If they theorize, they do so

> In regions mild, of calm and serene air,
> Above the smoke and stir of this dim spot
> Which men call earth.[56]

Women were (by courtesy) free of the morally ambiguous appetite for controversy which characterized professional science. The quotation from *Comus*, with which Whewell clinches his argument, echoes a sonnet he had addressed to Somerville in praise of her earlier work *The Mechanism of the Heavens*, in which he associated Somerville's gender with her freedom from factionalism or professional jealousy. The sonnet opens with an allusion to Mary's sex—'Lady'—and its closing lines describe her work as both intellectually and morally pure:

> thus we find
> That dark to you seems bright, perplexed seems plain,
> Seen in the depths of a pellucid mind,
> Full of clear thought, pure from the ill and vain
> That cloud the inward light![57]

[54] Richard Yeo, *Defining Science: William Whewell, Natural Knowledge, and Public Debate in Early Victorian Britain* (Cambridge: CUP, 1993), p. 84. Yeo quotes a letter from Lockhart to Whewell of 24 January 1834.

[55] Yeo, *Defining Science*, pp. 87, 89.

[56] Whewell, [unsigned review], 65, quoting Milton's *Comus* ('A Mask Presented at Ludlow-Castle, 1634', in *The Works of John Milton*, ed. Frank Allen Patterson, and others, 18 vols (New York: Columbia University Press, 1931–8), I (1931), pp. 85–123 (p. 85, l. 4)).

[57] Quoted in Mary Somerville, *Personal Recollections, from Early Life to Old Age, of Mary Somerville*, ed. Martha Somerville (London: John Murray, 1873), p. 172.

It is easy to see how women's supposed purity of mind is conflated with a 'lady''s supposed purity of character, rendering her both able and unable to participate in the uncertainties and controversies of scientific investigation. More interesting, though, is Whewell's argument that for women thought is usually a kind of seeing, a habit of mind which necessarily marginalizes them in a physical science in which prestige is increasingly invested in reasoning over observing.

Beyond the backhanded compliment he delivers to Somerville's sex, Whewell uses the bulk of his lengthy and authoritative review to deprecate the movement toward specialization in the sciences, and to mourn the growing division of the sciences and arts. Whewell agreed with Somerville's premise that the sciences are really one, but recognized that at present the scientific disciplines were involved in increasing specialization and subdivision. But where Somerville saw the direction of contemporary science as tending (at least in some areas, such as electricity and magnetism) towards rapprochement and unification, Whewell saw the opposite: a movement towards separation and mutual incomprehension. Boundaries were being formed and policed, separating not only disciplines but also sensibilities and mental capabilities:

the separation of sympathies and intellectual habits has ended in a destruction, on each side, of that mental discipline which leads to success in the other province. But the disintegration goes on, like that of a great empire falling to pieces; physical science itself is endlessly subdivided, and the subdivisions insulated.[58]

Somerville's book, for him, represented a contradiction of this literally small-minded, territorial practice. Urging against territoriality, he nonetheless adopted agricultural metaphors, summing up his complaint: 'the inconveniences of this division of the soil of science into infinitely small allotments have been often felt and complained of'.[59]

Whewell's references to land and its distribution do complex political work in the review. The suggestion might seem to be that, in Oersted's phrase, even 'the most ignorant among us [are] participators in a common inheritance', and that parcelling-up common intellectual assets by a minority is an injustice.[60] But since most of the reviews Whewell wrote during the 1830s were published in conservative journals (*The*

---

[58] Whewell, [unsigned review], 59.   [59] Ibid., 60.
[60] Oersted, 'The Comprehension of Nature by Thought and Imagination', p. 43.

British Critic and the *Quarterly Review*), it seems unlikely that the submerged political analogies in his spatial metaphors alluded to any democratization of property-holding. His regretful allusion to a disintegrating empire suggests, rather, that ownership, once established, is to be defended. Three years later, in the introduction to his *History of the Inductive Sciences*, Whewell wrote that a knowledge of scientific history is important because 'the examination of the steps by which our ancestors acquired our intellectual estate, may make us acquainted with our expectations as well as our possessions'.[61] Scientific knowledge is a kind of property that we inherit, and one that preserves and conserves its past.

This was an argument that Somerville might have supported. In her enormously popular 1848 book *Physical Geography*, she argued that 'the love of a country life', which contributed to making the British so successful in the outside world and in self-improvement, 'is chiefly owing to the law of primogeniture, by which the head of a family is secured in the possession and transmission of his undivided estate, and therefore each generation takes a pride and pleasure in adorning the home of their forefathers'.[62] The British system of land ownership, Somerville thought, had an effect on the British temperament and hence on Britain's fortunes in the world.[63] This is a literal version of Whewell's metaphorical comparison between land holding and intellectual advance, and both writers emphasize the role of property rights in defining, preserving, and promoting progress. Whewell goes on to stress that science progresses not by revolutions but by 'a series of developments', emphasizing the stability and orderliness which make the property system workable.[64]

[61] William Whewell, *History of the Inductive Sciences, from the Earliest to the Present Times*, 3 vols (London: Parker; and Cambridge: Deighton, 1837), I, p. 5.

[62] Somerville, *Physical Geography*, I, p. 263.

[63] The argument about the role of primogeniture in contributing to British national fortunes was an important one in political economy; nearly thirty years previously Malthus had argued that primogeniture had, 'for all practical purposes, annihilated the distinctions founded on rank and birth, and opened the fairest *arena* for the contests of personal merit in all the avenues to wealth and honours', and extended his point (as Somerville does) to the moral as well as the political basis of the nation by suggesting that 'the obligation generally imposed upon younger sons to be the founders of their own fortunes, has infused a greater degree of energy and activity into professional and commercial exertions than would have taken place if property in land had been more equally divided' (*Principles of Political Economy*, I, p. 436). Thus Malthus argues that inequality of land division promotes equality of opportunity.

[64] Whewell, *History of the Inductive Sciences*, I, p. 10.

So in his review of Somerville's book, far from urging a radical agenda of the opening up of the ownership of knowledge, Whewell was, rather, proposing a sort of enclosure movement in science; as with the enclosure acts which had been changing the English landscape and agricultural life for two centuries or more, the landscape of science must be reshaped and reformed so as to avoid the endless production of boundary lines. The scientific disciplines must be combined into a single more efficient estate.

The improvement of the countryside along picturesque and efficient lines was a frequently-discussed topic in the *Quarterly Review*, the journal in which Whewell's review appeared, during the first decades of the nineteenth century.[65] This attention was in turn part of a far wider cultural debate about the English landscape, in which political ideology and aesthetic fashion claimed almost equal authority. Metaphors referring to enclosure and the physical landscape invoked political as well as aesthetic aspects of land use. Similarly, as we saw in the first chapter, discourse could move rapidly between metaphors of 'real' landscapes and metaphors of represented ones, particularly landscapes in paintings. In his review of Somerville's book, for instance, Whewell draws an analogy between techniques of landscape art and of scientific explanation:

The long-drawn vista, the level sunbeams, the shining ocean, spreading among ships and palaces, woods and mountains, may make [a] painting offer to the eye a noble expanse magnificently occupied; while, even in the foreground, we cannot distinguish whether it is a broken column or a sleeping shepherd which lies on the earth, and at a little distance we may mistake the flowing sleeve of a wood-nymph for an arm of the sea. In like manner, language may be so employed that it shall present to us science as an extensive and splendid prospect, in which we see the relative positions and bearings of many parts, though we do not trace any portion into exact detail—though we do not obtain from it precise notions of optical phenomena, or molecular actions.[66]

---

[65] See for instance *Quarterly Review*, 15 (April 1816), 187–235, for a review of *Reports of the Society for Bettering the Condition of the Poor* and others, especially p. 217 on the benefits of dividing farms; 24 (January 1821), 400–19, review of *Transactions of the Horticultural Society of London* and *Memoirs of the Caledonian Horticultural Society*, especially pp. 413–14 on the superiority of British horticultural products; 33 (March 1826), 455–73, review of *Minutes of Evidence taken before the Select Committee of the House of Lords appointed to inquire into the State of Ireland* and *Minutes of Evidence taken before the Select Committee of the House of Commons appointed to inquire into the State of Ireland*, especially pp. 456–7 and p. 465 on the evils of minute subletting.

[66] Whewell, [unsigned review], 55.

In a rhetorical manoeuvre frequently found in both picturesque art and non-fiction prose of the period, the reader is placed on a high vantage point from which he or she is invited to regard the 'splendid prospect' of the landscape. It was one of Whewell's favourite motifs. In the *History of the Inductive Sciences* he uses it again to make the same point, but this time it is powered by a reference to Moses leading the Jews out of exile. A historical survey of science

> may be instructive as well as agreeable; it may bring before the reader the present form and extent, the future hopes and prospects of science, as well as its past progress. The eminence on which we stand may enable us to see the land of promise, as well as the wilderness through which we have passed.[67]

From high up we can see the past and the future; we are, in a sense, inspired. The trope of the raised vantage point was a cliché of topographical art and had long been put to political use as part of the visual rhetoric of conquest and ownership associated with both written and visual accounts of landscape in the New World. It was, for instance, commonly used in exploration narratives where the writer sought to impress upon the reader the authority of his claim to the land spread out below him.[68] Whewell's rhetoric in his review of Somerville's book is concerned not only with the present but with the future scientific landscape. He writes that science can be presented to the reader as 'an extensive and splendid prospect, in which we see the relative positions and bearings of many parts, though we do not trace any portion into exact detail'. He puns on 'prospect', referring primarily to a landscape that we see laid out before us but at the same time suggesting a view into the future. This double sense of 'prospect' is an example of the way in which a spatial metaphor can imply a temporal meaning. ('Prospect' in its mining-related senses did not enter the language as a noun until five years after this review, and did not exist as a verb in this context until mid-century; these new usages would link the geographic and the predictive senses of the word in dramatic fashion.) Whewell wants his reader to believe that to understand the whole present state of science is to be able to predict its future development.

---

[67] Whewell, *History of the Inductive Sciences*, I, p. 4–5.

[68] On the use of this trope in exploration narratives, see Bruce Greenfield, 'The Problem of the Discoverer's Authority in Lewis and Clark's *History*', in *Macropolitics of Nineteenth-Century Literature: Nationalism, Exoticism, Imperialism*, ed. Jonathan Arac and Harriet Ritvo (Philadelphia: University of Pennsylvania Press, 1991), pp. 12–36 (pp. 28–9).

Prospects are one spatial trope Whewell uses to describe the current state and future aspirations of science. 'Bearings' are another: 'we see the relative positions and bearings of many parts', ' the bearing of new facts in one subject upon theory in another'. Especially when allied with 'positions', 'bearings' introduces the semantic field of navigation into Whewell's rhetoric, recalling that this review was written during the onset of the heyday of British domination of the world's seas and at a time when a considerable portion of British naval power was being used for exploration rather than hostilities. 'Bearings' and 'positions' make an analogy between scientific investigation and the process of geographical exploration. This analogy in turn depended on the inference that the logic of mathematical navigation developed by the European seafaring traditions and formalized in European science applied not only within the system in which it was formed but could also be successfully projected to apply to parts of the globe as yet unknown. In other words, European scientific spatial models for ordering information (such as the location of one place in relation to another) were taken to be effective when used outside Europe in very much the same way that scientific reasoning about the visible part of the universe could be taken to be effective also about the unseen parts. Just as global navigation was possible because the space of the earth is homogeneous, inductive and predictive science was legitimate because the universe itself was homogeneous.

When 'bearings' and 'positions' appear in Whewell's rhetoric they refer to an immensely ambitious and totalizing kind of science. 'Bearings' and 'positions' in non-specialist discourse carried allusions to the systematized, systematizing bodies of knowledge such as longitude and latitude that made large-scale exploration and an imperializing project possible. And this is exactly the kind of science Whewell was advocating: a science free of the clutter of disciplinary borders and wide open to a totalizing future.

Querying the number and effect of disciplinary boundaries was not new to the nineteenth century:

In the wide field of intelligibles, appear some parts which have been more cultivated than the rest; chiefly on account of the richness of the soil, and its easy tillage; but partly too, by reason of the skilful and industrious hands under which it has fallen. These spots, regularly laid out, and conveniently circumscribed, and fenced round, make what we call the *Arts and Sciences*: and to these have the labours and endeavours of the men of curiosity and learning, in all ages, been chiefly confined. Their bounds have been enlarged from time to time, and new acquisitions made from the adjoining waste; but still the space of ground they possess is but narrow; and there is room either to extend them vastly; or to lay out new ones.[69]

Ephraim Chambers's *Cyclopaedia; or An Universal Dictionary of Arts and Sciences* appeared early in the eighteenth century and went into many editions through the next hundred years. Chambers's 'Preface' uses a great number of colourful spatial metaphors to account for the formation of disciplinary boundaries within the 'wide field of intelligibles'. After describing the arts and sciences in terms of plots of land, Chambers goes on to put his spatial metaphors to more critical use, commenting on the sciences in particular that:

They were divided, by their first discoverers, into a number of subordinate provinces, under distinct names; and have thus remained for time immemorial, with little alteration. And yet this distribution of the land of science, like that of the face of the earth and heavens, is wholly arbitrary; and might be altered, perhaps not without advantage. Had not Alexander, Caesar, and Gengiskan lived, the division of the terraqueous globe had, doubtless, been very different from what we now find it: and the case would have been the same with the world of learning, had no such person been born as Aristotle. The first divisions of knowledge, were as scanty and ill concerted, as those of the first geographers; and for the like reason: and though future Bacons, Cartes's, and Newtons, by opening new tracts, have carried our knowledge a great way further; yet the regard we bear to the antient adventurers, and the established division, has made us take up with it, under all its inconveniences, and strain and stretch things, to make our later discoveries quadrate thereto. I do not know whether it might not be more for the general interest of learning, to have the partitions thrown down, and the whole laid in common again, under one undistinguished name.[70]

Ephraim Chambers, like Whewell a century later, calls for a single, open science, and his account of the development of the disciplines as they stood at the time of the *Cyclopedia* is intensely and richly spatial. The fluid play on the senses of 'field', from farming ('the richness of the soil, and its easy tillage'), to battle (allusions to Alexander, Caesar, and Genghis Khan), returning again to agriculture in the suggestion that science be 'laid in common again', imbues the notion of disciplinarity with a confusion of conflicting associations. Knowledge is a territory to

---

[69] E[phraim] Chambers, *Cyclopaedia; or An Universal Dictionary of Arts and Sciences; […]*, 2nd edn, corrected and amended, with additions, 2 vols (London: Midwinter and others, 1738), I, p. ix.
[70] Ibid., I, p. ix.

be both conquered and cultivated, and those who deal in knowledge are both aggressors and nurturers.

Compare Chambers's divided metaphorical landscape with the section of Whewell's review of Somerville where he indicts the break-up of the 'empire' of science into disciplines and sub-disciplines:

> If a moralist, like Hobbes, ventures into the domain of mathematics, or a poet, like Goethe, wanders into the fields of experimental science, he is received with contradiction and contempt; and, in truth, he generally makes his incursions with small advantage, for the separation of sympathies and intellectual habits has ended in a destruction, on each side, of that mental discipline which leads to success in the other province. But the disintegration goes on, like that of a great empire falling to pieces; physical science itself is endlessly subdivided, and the subdivisions insulated. The inconveniences of this division of the soil of science into infinitely small allotments have been often felt and complained of.[71]

Here again the tensions between agricultural and military senses of 'field' are played out, with the pastoral wanderings of the poet quickly turning into raiding parties or incursions. The anthropologist James Clifford has suggested that various aspects of the 'field' metaphor are traditionally coded as masculine (senses involving 'agriculture, property, combat') and others as feminine (involving 'penetration, exploration and improvement').[72] Whewell is certainly interested here in Somerville's gender and its effect on her book, which may have contributed to his drawing 'field' metaphors from what Clifford sees as the 'masculine' range. But despite this concentration on just a few sources of metaphor, the writing cannot settle on one register or one tone. Whewell's image of the great empire in disintegration is raw and powerful; but this rawness is traduced by the urbanity of the final sentence, where regret dwindles into mere irritation caused by 'inconveniences'.

The tone is unstable, partly because throughout the review reference teeters between the concrete and the conceptual or metaphorical, as well as between different levels of metaphor. Somerville's ambitious and successful cross-disciplinary project engenders in Whewell an uncertainty about the powers of verbal language which causes him to return insistently but unevenly to extended metaphors of spatial location, using them sometimes apparently consciously, sometimes not, and frequently

---

[71] Whewell, [unsigned review], 59.
[72] James Clifford, 'Notes on (field)notes', in *Fieldnotes: The Makings of Anthropology*, ed. Roger Sanjek (Ithaca, N.Y.: Cornell University Press, 1990), pp. 47–70 (p. 65).

confusing the reader by juxtaposing contradictory versions of related metaphors. Whewell often commented on the difficulty of using verbal language to write about spatial concepts, and his use of spatial metaphors here attests to those difficulties. Despite his standing as a philologist, Whewell insisted that the common language of all the sciences was not verbal at all: 'mathematical language is the very plainest English which can be used to those that can understand it'.[73] He found complex spatial relations particularly unsuited to expression in words: in his *Philosophy of the Inductive Sciences* he notes that mineralogy has given rise to 'an extensive and recondite branch of mathematical science' because 'the relations of space which are involved in the forms of crystalline bodies, though perfectly definite, are so complex and numerous, that they cannot be expressed, except in the language of mathematics'.[74]

Science which cannot be expressed verbally but which requires an advanced knowledge of mathematics is necessarily a science accessible to only a small fraction of the population. Whewell, for all his admiration of Somerville's book, is not really in sympathy with its popularizing project. One of the passages that Whewell particularly enjoys is Somerville's account of the great alarm manifested by the Parisian public when they became aware of the impending approach of a comet:

It appeared that the comet would cross the earth's orbit; what mischief might not come of this? It was true that the earth would not be near the crossing at that time; but then, might not the orbit itself be seriously injured? Instead of an imaginary line in the trackless ocean of space, the fears of our friends appear to have represented to them the earth's orbit as a sort of railroad, which might be so damaged [...] that the earth must stick or run off, the next time the revolving seasons brought her to the fatal place.[75]

This passage brings together a number of questions about borders in early nineteenth-century intellectual life. First, national borders: Whewell's account of this incident is distinctly xenophobic in tone, taking pleasure in the discomfiture and ignorance of the French populace. The ironic phrase 'our friends' only exaggerates Whewell's comic disdain for the French. (It has been suggested that despite Whewell's wide travel

[73] Simon Schaffer, 'The History and Geography of the Intellectual World: Whewell's Politics of Language', in Menachem Fisch and Simon Schaffer eds, *William Whewell: A Composite Portrait* (Oxford: Clarendon Press, 1991), pp. 201–31 (p. 225), quoting a letter from Whewell to Lady Murchison, 19 November 1841.

[74] William Whewell, *The Philosophy of the Inductive Sciences, Founded Upon Their History*, 2 vols (London: Parker, 1840), I, lxiv.

[75] Whewell, [unsigned review], 58.

experience, he took a particularly nationalist and narrow view of European science.[76]) Second, the passage describes confusion between different levels of materiality and metaphoricity in thinking about borders: the Parisians wrongly consider an orbit to be material, a physical boundary in space. For the mob it exists as a tangible object; for the educated it is a notional line, a geometrical rather than physical entity. Third, there is the matter of borders intended to keep out the uninitiated. Whewell distinguishes forcefully between the educated and the ignorant, those with access to knowledge and those without. Much as his review argues for a relaxation of boundaries between scientific disciplines, and a liberalization of the opportunities for people trained in one branch of science (or other types of knowledge) to move about in another branch, Whewell asserts a number of shadowy but nonetheless powerful boundaries of his own in this piece—between the ignorant mob and the amused elite, and between male and female minds.

John Gibson Lockhart, the editor of the *Quarterly Review*, noticed the boundary-creating tendency in Whewell's reviews of the 1830s. He described Whewell, together with the optics specialist David Brewster and the geologist Charles Lyell, as 'heavy, clumsy performers; all mere professors, hot about little detached controversies, but incapable of carrying the world with them in large comprehensive *resumés* of the actual progress achieved by the combined efforts of themselves and all their rivals'.[77] Ironically, the fault of which Lockhart was accusing Whewell—increasing rather than reducing the boundary-laden nature of science—was the very one Whewell so much deprecated in the work of his contemporaries.

And yet it was Whewell who, at a meeting of the British Association, coined the word 'scientist', a word designed to include practitioners from all the scientific disciplines and sub-disciplines and which thus performs exactly the inclusive work he advocated in his review of Mary Somerville's project.[78] Whewell's role as critic and supporter of boundaries within the field of science is a particularly fruitful case study because of the eloquence of his writing and because of his ubiquity and influence in early nineteenth-century British science. But he was not exceptional in promoting either the belief that the universe was

[76] Maurice P. Crosland, *Historical Studies in the Language of Chemistry* (London: Heinemann, 1962), p. 279.
[77] Lockhart to John Murray, 26 October 1839, quoted in Yeo, *Defining Science*, p. 84.
[78] Appropriately, 'scientist' first appeared in print in Whewell's review of Somerville's *On the Connexion* (p. 59).

ultimately one homogeneous field or that science must recognize some boundaries (such as gender or national ones) while doing its utmost to minimize others (such as disciplinary boundaries). Using a rhetoric based on several kinds of spatial division (landholding, border disputes, and imperial fragmentation) allowed Whewell to discuss the politics of science from the point of view of someone who is afraid he may soon no longer be able to see the wood for the trees.

4

Space and the Languages of Science

I have been a long time out of England, you know, and that
must be my excuse [...] but your provincial superior people, who
have attained a small reputation—oh, those are terrible! I have
just been listening to one of them, who was speaking to C—the
geologist, and propounding his 'Asses' Bridge' as if it were a new
revelation. I left him in despair [...]. [1]

The scientist and industrial entrepreneur John Bryant uses this attack
on the insularity of British society to woo Alice Helmsby at a soirée
celebrating a meeting of the British Association for the Advancement of
Science in Geraldine Jewsbury's 1848 novel *The Half Sisters*. The BA's
annual meeting moved to a different city each year and a great deal
of store was set within the organization on the connections with local
people which this migration was designed to promote.[2] Bryant seems to
be quite out of sympathy with this ideal, however; he uses a rapid series
of shifts from national identity to regional identity and class to distance
himself and his fellow scientists from the BA's hosts. And the narrator
supports Bryant's view of the matter; though the scientists and the locals
mingle at the soirée, they fail to communicate at anything but the most
superficial level. It is as if they are speaking different languages. The
'*savans*' (the term highlights their linguistic difference from the English
provincials)

---

[1] Geraldine Jewsbury, *The Half Sisters* (1848) (Oxford: Oxford World's Classics,
1994), p. 55 (ch. 10).
   [2] As Jack Morrell and Arnold Thackray write, 'steady travel across the length and
breadth of the United Kingdom was one obvious means by which the Association
fostered cohesion among its many constituencies and a shared national sense of the values
of science and of Empire (*Gentlemen of Science: Early Years of the British Association for
the Advancement of Science* (Oxford: Clarendon Press, 1981), p. 108).

blandly attempted to give intelligible replies to questions (put by those lucky enough to get an introduction) which required the knowledge of a life-time to be compressed into a portable answer, that might be carried away like a receipt for venison sauce.[3]

Bryant's experience overseas (he has been away for years, 'establishing iron works somewhere, where hardly any body has ever travelled even') allows Jewsbury to present the inadequacy of the BA's audience as characteristically English.[4] The locals are starstruck but depressingly unelevated, impressed by the celebrity of the scientists but ignorant of the real dignity of science.

Jewsbury's novel enters into an existing debate about the advantages and disadvantages to science of the English national character and social attitudes. The debate had run for the best part of two decades and had often been heated in its comparisons of English with continental science and scientific institutions. One of the opening salvos of the debate, and one of the most polemical, was Charles Babbage's attack on the complacent British scientific world in his 1830 *Reflections on the Decline of Science in England*:

It cannot have escaped the attention of those, whose acquirements enable them to judge, and who have had opportunities of examining the state of science in other countries, that in England, particularly with respect to the more difficult and abstract sciences, we are much below other nations, not merely of equal rank, but below several even of inferior power.[5]

Even as undergraduates at Cambridge, Babbage and his contemporary John Herschel had been aware of the dangers of British scientific insularity. With a handful of others in 1812 they founded the Analytical Society, which aimed to promote the British adoption of the more advanced mathematical methods and the differential notation in use on the Continent. Within a decade they had won the argument within the university, and from Cambridge the use of analytical methods spread throughout the country.[6] But this victory did not of course affect all aspects of British science; modernization was still essential, they felt, in

[3] Jewsbury, *The Half Sisters*, p. 53 (ch. 10). Bryant turns the tables on the locals by carrying Alice herself away when he marries her.

[4] Ibid., pp. 49–50 (ch. 10).

[5] Charles Babbage, *Reflections on the Decline of Science in England* (London: Fellowes and Booth, 1830; repr. Farnborough: Gregg, 1969), p. 1.

[6] W. W. Rouse Ball, *A History of the Study of Mathematics at Cambridge* (Cambridge: CUP, 1889), pp. 120, 123.

some areas. Babbage's *Reflections on the Decline of Science in England*, quotes Herschel writing in uncharacteristically pessimistic vein that with regard to some of the sciences 'it is vain to conceal the melancholy truth. We are fast dropping behind. In mathematics we have long since drawn the rein, and given over a hopeless race. In chemistry the case is not much better.'[7]

This falling behind in British science was, for Babbage, partly the result of the maintenance of outmoded boundaries within the scientific establishment. Class privilege, instead of learning and achievement, dominated the scientific elite, particularly that of the Royal Society, which—in the absence of systematic modern scientific curricula in the universities—was among the chief arbiters of British science. A widespread failure to read foreign languages meant that British scientists were often not up to date with contemporary European discoveries. And British science lacked the systems of peer review developing in continental practice. Babbage argued that a kind of lazy insularity was to blame for the deceleration of British achievement. He, Herschel, and others offered possible solutions to this state of affairs; put together these reforms would promote the creation of a meritocratic, professional, outward-looking scientific establishment in Britain, one better organized and better connected than before. In part, this was to be achieved through a new commitment to internationalism.

The early decades of the nineteenth century, even during the immense European hostilities up till 1815, had already seen an increasing emphasis on internationalism in scientific research. Napoleon had been a great promoter of science and had even granted a passport for Sir Humphry Davy and his retinue (including, in a fairly humble capacity, the young Michael Faraday) to travel in Continental Europe in 1813 at the height of the Anglo-French wars. After the fall of Napoleon and the Congress of Vienna the recognition of the international nature of scientific endeavour combined with renewed opportunities for travel to remind British science of the disadvantages of isolation. Changes in the scientific establishment in Britain manifested this awareness. From within a couple of years of its inception, for example, the British Association for the Advancement of Science was a major force in the broadening of British science's European and American links. The British Association quickly made the participation of foreign scientists in their meetings a priority. By the BA's tenth meeting, held in Glasgow

---

[7] Quoted in Babbage, *Reflections on the Decline of Science in England*, p. viii.

in 1840, this international participation was well established. Of the twenty-two toasts delivered at an official dinner during the Glasgow meeting, five were in honour of individuals and scientific societies from overseas, and a sixth celebrated technological internationalism: 'railway communication, & other improvements which tend to facilitate intercourse between mankind, and thereby promote friendly relations'.[8] These were not merely practical points: internationalism could be, and was, mobilized to advance the ideology of the disinterested progressivism of science. As Jack Morrell and Arnold Thackray have shown in their study of the early years of the British Association, 'the idea of science as something above party politics and sectarian dispute was easily extended to the notion that science was an agent of nothing less than universal philanthropy'.[9]

This internationalist ideology was easily combined with scientific triumphalism. Thomas Dick, whose lengthy and excitable book *On the Improvement of Society by the Diffusion of Knowledge* was published in 1833, felt that the development of Mechanics' Institutes and other public educational projects, the expansion of cheap publications, and what he saw as the widespread desire to learn, were contributing to an end to the injustice and ill-will which had previously led to European conflict:

These circumstances, notwithstanding some gloomy appearances in the political horizon, may be considered as so many preludes of a new and happier world; when intellectual light shall be diffused among all ranks, and in every region of the globe, when Peace shall extend her empire over the world;—when men of all nations, at present separated from each other by the effects of ignorance, and of political jealousies, shall be united by the bonds of love, of reason, and intelligence, and conduct themselves as rational and immortal beings.[10]

The millenarian element in Dick's view of science stemmed from his evangelicalism.[11] But the idea that the diffusion of science and

---

[8] Morrell and Thackray, *Gentlemen of Science*, plate 23.   [9] Ibid., p. 373.

[10] Thomas Dick, *On the Improvement of Society by the Diffusion of Knowledge: or, An Illustration of the Advantages which would Result from a More General Dissemination of Rational and Scientific Information Among All Ranks* (Edinburgh: Waugh and Innes; Dublin: Curry; London: Whittaker, 1833), p. 16.

[11] William J. Astore, whose book *Observing God: Thomas Dick, Evangelicalism, and Popular Science in Victorian Britain and America* (Aldershot: Ashgate, 2001) reveals Dick as an interesting and influential figure in the relationship of science and religion in the early nineteenth century, argues that Dick 'intended that his writings would teach moral and didactic lessons, which would serve both to advance the millennium and as preparation for future explorations of God's works in the afterlife' (p. 6).

technological progress was bringing on a new dawn for European politics was not confined to millenarianists. A decade and a half later, and with characteristic optimism, the great scientific educator Mary Somerville argued in the year of European revolutions that technological advances were promoting intercultural exchange and ushering in a new age of international unity:

The history of former ages exhibits nothing to be compared with the mental activity of the present. Steam, which annihilates time and space, fills mankind with schemes for advantage or defence; but however mercenary the motives for enterprise may be, it is instrumental in bringing nations together, and uniting them in mutual bonds of friendship. The facility of communication is rapidly assimilating national character. Society in most of the capitals is formed on the same model; individuality is only met with in the provinces, and every well educated person now speaks more than one of the modern languages.[12]

Somerville's own scientific career had always had an international flavour. Her first work had been a version in English of the French astronomer Laplace's monumental *Mécanique céleste*. She had lived in Italy since 1838 and her European connections had been put to use in giving the 1833 BA meeting some of its internationalism.[13] Perhaps her own cosmopolitan scientific career led her to underestimate the growing nationalism of European politics. However that may be, her opinion about the ease of international communication was certainly premature. The question of the languages of science became a matter of considerable urgency as papers dealing with major scientific developments were circulated in Britain, France, Scandinavia, Germany, and Italy. Over the next few decades the language of many of the physical sciences would come to be dominated by mathematics rather than inexact verbal expressions; but in the early nineteenth century the bulk of scientific research was still presented in words, and though the primary audience for scientific publications was now scientists themselves, practitioners still expected their work to be reviewed in and disseminated partly via a literary culture aimed at an educated general reader. Under these circumstances, the problem of the boundaries imposed on scientific work by national languages could not be ignored.

The tradition of scholarly humanism which was implied in verbal, rather than mathematical, scientific language carried with it a strong

---

[12] Mary Somerville, *Physical Geography*, 2 vols (London: John Murray, 1848), II, p. 274.

[13] Morrell and Thackray, *Gentlemen of Science*, p. 374.

loyalty to Latin, the universal language of medieval and Renaissance international scholarship. Even as late as 1820, Latin was sometimes used as a lingua franca for science: the Danish natural philosopher Hans Christian Oersted's seminal paper announcing his discovery of electromagnetism was published originally in Latin, as *Experimenta circa effectum conflictus electrici in acum magneticam*, before being translated quickly into a variety of European languages.[14] So debates regarding scientific language were twofold, encompassing both the choice between verbal and symbolic or mathematical language and also the future of communicating in an international arena using vernacular verbal languages.

At its most basic level, there was the problem of what names to give things. Gillian Beer notes that 'questions of nomenclature in the later nineteenth century became connected with current movements in language theory and nationalist practice'.[15] Although in the early decades of the century the study of linguistics was not inflected in the same ways as during the heyday of Max Müller and his successors, nonetheless national and nationalist practices and prejudices were sometimes involved in the determination of nomenclature. We might once again take William Whewell as a case study in the instability of British science's relationship with languages. Whewell's authority in philological and philosophical matters was very great. He was called on by experts in a number of sciences to help with the baptism of newly discovered, invented or observed objects. And on his own behalf he was deeply embroiled in controversy over the development of a new nomenclature and notation for chemistry, a controversy which had a strongly international dimension.

The problem was that because of the enormous advances in chemical analysis since the late eighteenth century, and because of the development by John Dalton of a modern atomic theory which tried to explain chemical bonding and affinities and to make prediction possible, earlier systems for naming chemical substances and recording their interactions were hopelessly outdated. By the second quarter of the nineteenth century, several major notational systems were competing for European

---

[14] H. A. M. Sneiders, 'Oersted's Discovery of Electromagnetism', in Andrew Cunningham and Nicholas Jardine, eds, *Romanticism and the Sciences* (Cambridge: CUP, 1990), pp. 228–40 (p. 228).

[15] Gillian Beer, 'Translation or Transformation? The Relation of Literature and Science', in *Open Fields: Science in Cultural Encounter* (Oxford: Clarendon Press, 1996), pp. 173–95 (p. 89).

supremacy. One was Dalton's system, which used symbols to denote not only which atoms were involved in a substance but also how they were arranged. Dalton's sign for water (thought to consist of one atom of oxygen and just one of hydrogen) was 'O☉'. Another competing system was that of the Swedish chemist Jon-Jacob Berzelius, who similarly favoured a symbolic language, but one relying on a combination of letters for some of the elements and symbols for other elements and numbers of atoms: water, in Berzelius's notation, was written 'Ḣ', the dot over the H indicating a single atom of oxygen combining with the single atom of hydrogen. A third major contender was the algebraic system, which is the closest of the three to the present one: the early nineteenth century notation for water in this notation is 'HO'.[16]

Whewell hoped to tidy up the confusions and distracting contestations preventing chemistry from developing a single authoritative notation and language. In the early 1830s he advocated a system of nomenclature and classification combining verbal roots and numerical notation in a manner conformable with the grammar of algebra. The authority of this system was to be derived chiefly from the notion of the absolutely truthful representations of reality inherent in numbers, and, intriguingly, from reaching back into the history of natural science and reviving the partial use of Latin. His suggestion for the roots of the new notations for oxides of metals was that

The Latin name should be used, for the sake of European communication. Thus, we should have *fe* (*ferrum*) iron, *sn* (*stannum*) tin, *cu* (*cuprum*) copper, *ag* (*argentum*) silver, *au* (*aurum*) gold.[17]

Whewell's suggestion of using Latin as a language common to all European scientists was unfashionable. The vernacular had been established as the medium of scientific communication more than a century previously. Whewell's return to Latin was partly a concession to the successful rival practice of the great Swedish chemist: Whewell notes that 'these are the symbols commonly used by those who follow Berzelius'.[18] But reverting to a classical language was also an attempt to invoke on behalf of Whewell's system the authority of the centuries of

---

[16] These translations of 'water' are helpfully listed in the 1850 edition of Charles Daubeny's *An Introduction to the Atomic Theory* (Oxford: University Press, 1850), pp. 110–11.

[17] William Whewell, 'On the Employment of Notation in Chemistry', *Journal of the Royal Institution of Great Britain*, 1 (1830–1), 437–53 (p. 447).

[18] Ibid., p. 447.

Latin communication between natural philosophers, and of the 'father-tongue' whose demise as a living language is the condition of its definitive power. Latin is an effective language for making claims to authority in the world of learning because—among other things—it erects a boundary between those with a privileged educational background and those without. Tying his language of symbols to Latin could be presented as a universalizing move, but also in fact gave it the exclusivity which constituted part of professional acceptance.

In this paper of 1830–1, Whewell engaged in direct argument with Berzelius over his proposed modifications to Berzelius's symbolic system. Both sides expressed themselves in nationalist terms. Whewell's paper made hostile remarks about 'the gross anomalies and defects with which the foreign notation is disfigured' and explicitly linked the success of this notation with Berzelius's high personal prestige so that an insult to the system was a insult to the man: the Swede

has lent the weight of his great authority to a system which [...] violates mathematical propriety so entirely, that it must always be disagreeable to see an example of it, for any person who has acquired the first rudiments of algebra [...]. [19]

Throughout his paper Whewell takes a patronizing tone towards continental scientists, concluding with the pompous remark that 'I should hope that what I have said may tend to induce our chemists to purify and improve the foreign system, before it is admitted to a familiarity and circulation among us'.[20]

Whewell's paper was published with Michael Faraday's assistance in the *Journal of the Royal Institution*.[21] The combination of criticism of Berzelius's system with this smugly superior tone irritated Berzelius into responding in similarly nationalist terms:

Die Gelehrten dieser Nation, wenig bekannt mit fremden Sprachen, bekommen erst spät Kenntniss von den Fortschritten der Wissenschaft in anderen Ländern, und finden stets Vorwände zur Vertheidigung dieser Trägheit, unter welchen nicht selten der verkommt, dass es foreign (auslandisch) ist, was dann entweder

---

[19] Ibid., pp. 437, 441.   [20] Ibid., p. 453.
[21] See Faraday to Whewell, 21 February 1831, in Frank A. J. L. James, ed., *The Correspondence of Michael Faraday* (London: Institution of Electrical Engineers, 1991–), I, pp. 549–50, in which Faraday mentions his assistance in seeing the excerpt through the press and concurs with Whewell's wish to see fewer systems used in chemistry.

höchst wichtig oder schon etwas alt sein muss, um allgemeinere Aufmerksamkeit
zu gewinnen.[22]

[The scholars of this nation, [England] who are little acquainted with foreign
languages, receive late knowledge of the advances of science in other countries,
and continually find pretexts for the defence of this laziness, so that if something
is foreign it must be extremely important or else already rather old in order to
gain attention.]

Berzelius's aggression was directed against exactly the kind of insular attitude which Babbage and Herschel had identified as seriously damaging
the prospects of British science and which Whewell, despite his travels
on the Continent, seemed—at least in his battle with Berzelius—to
endorse.

Even where there was no question of rivalry or aggression, national
character and national language were sometimes thought to affect the
scientific productions of the different countries of Europe. In the late
1840s the great Prussian scientific traveller and writer Alexander von
Humboldt published his multi-volume work *Kosmos*, a monumental
account of the nature of the solar system, the globe, its geology and
geography, and a history of beliefs about these topics since the ancient
world. Mary Somerville, who was about to publish her much more
limited book *Physical Geography*, read Humboldt's work and 'at once
determined to put my manuscript in the fire', but was persuaded
by Herschel to publish regardless.[23] A few years later she received a
letter from Humboldt praising her book warmly and contrasting their
approaches to the subject. He conceived the difference in national
terms. His *Cosmos* was 'an essentially German production'. 'There is,
I know, much good will for my Cosmos in your country; but there
are *forms* of literary composition, as in the variety of the races and
the primitive difference of languages.'[24] German science used different
forms because Germans were a different people and spoke a different
language. Humboldt made this argument several times in *Cosmos*
itself:

[22] Jacob Berzelius, *Jahres-Bericht über die Fortschritte der physichen Wissenschaften von
Jacob Berzelius: Eingereicht an die schwedische Akademie der Wissenschaften, den 31. März
1832*, trans. F. Wöhler, 12 (Tubingen: Laupp, 1833), 168–70 (p. 168).
[23] Mary Somerville, *Personal Recollections, from Early Life to Old Age, of Mary
Somerville*, ed. Martha Somerville (London: John Murray, 1873), p. 286.
[24] Ibid., p. 288.

Thought and language are of old intimately allied: if the language employed lends to the presentation grace and clearness, if by its organic structure, its richness, and happy flexibility, it favours the attempt to delineate the phaenomena of nature, it at the same time reacts almost insensibly on thought itself, and breathes over it an animating influence. *Words*, therefore, are more than signs and forms; and their mysterious and beneficent influence is there most powerfully manifested, where the language has sprung spontaneously from the minds of the people, and is on its own native soil. Proud of my country, whose intellectual unity is the firm foundation of every manifestation of her power, I look with joy to these privileges of my native land.[25]

Where Somerville had stressed the growing homogeneity of European culture, including scientific culture, Humboldt retained the German Romantic idea that though the truths of science belonged to all nations, national character was still present in the expression of those truths. He associated the German national character strongly with liberal political ideals (sometimes in the face of the actual politics of contemporary Prussia). It seemed to him, for example, to have 'deep meaning' that the German phrase for 'in the open air', the setting for his kind of scientific investigation, was '*in das Freie*', a phrase with clear ideological connotations that spilled over into his conception of natural science.[26]

Differences in national character and hence in national scientific practice were a hot topic among scientific commentators during the 1830s. Babbage's blast of 1830 provoked a furore which was quickly inflamed by the anonymously published defence of British science, *On the Alleged Decline of Science in England, By a Foreigner*, which appeared the following year. This book was introduced with a foreword by Michael Faraday, in which—characteristically—Faraday committed himself to neither side of the argument, though he did admit that it was 'an extraordinary circumstance for English character to be attacked by natives and defended by foreigners'.[27]

Utterly rejecting Babbage's claim that Bonaparte had been a great promoter of science, and arguing that the French Revolutionary valorization of abstract mathematics had thrown French science off balance by retarding applied work, *On the Alleged Decline* clearly had a subtext

---

[25] Alexander von Humboldt, *Cosmos: Sketch of a Physical Description of the Universe*, trans. under the superintendence of Edward Sabine, new edn, 5 vols (London: Longman and John Murray, 1856), I, p. 41.

[26] Ibid., I, p. 6.

[27] [Gerard Moll], *On the Alleged Decline of Science in England, By a Foreigner* (London: Boosey, 1831), n. p.

based in contemporary European national politics. The author was the Dutch scientist Gerard Moll, a friend of Faraday's. Though Moll argued that in England the scientific world was more meritocratic than in continental Europe, he admitted that there was a major obstacle preventing British scientists keeping up with European advances, and that was language.

Since the Latin has ceased, very happily in many respects, to be the common medium of intercourse of the learned in every country, the scientific intercourse between different nations has become cramped by the necessity of learning many foreign languages; at least three or four are indispensably necessary [...] but in England, the number of those who acquire a smattering of French is very small, and still smaller is the number of those who know enough of German to read a book in that language without considerable trouble.[28]

Heated debate about the validity of Moll's and Babbage's accounts of British science continued for over a decade, and not only within specialist fora. The London mathematician Augustus De Morgan published an essay on 'English Science' in the *British and Foreign Review* in 1835, asserting forthrightly that 'there never was a time when the [scientific] prospects of this country, in comparison with those of others, showed greater promise'.[29] On the other side of the argument, William Grove, recently elected to the Royal Society for his work on electricity, was able to insist in *Blackwood's Edinburgh Magazine* that 'in no civilized nation is [the] contempt of physical science carried to a greater extent than in England, the country of commerce and manufacture'.[30]

With Grove's barb about 'the country of commerce and manufacture' we return to the politics of class and the utility of knowledge. Grove had just been made Professor of Experimental Philosophy at the London Institution, a body founded to bring together science and commercial interests, challenging the existing elite scientific institutions. Clearly he had a good personal and professional reasons for attacking what he saw as a short-sighted and peculiarly English disdain for practical and applicable science, which was leaving Britain to fall behind her continental rivals.

Conceptions of Englishness and of class were deeply imbricated in contemporary arguments about disciplinarity and the practice of science.

---

[28] Ibid., pp. 7–8.
[29] [Augustus De Morgan], 'English Science', *British and Foreign Review* 1.1 (1835), 134–57 (p. 153).
[30] [William Grove], 'Physical Science in England', *Blackwood's Edinburgh Magazine*, 54 no. 336 (October 1843), 514–25 (p. 516).

For Moll, France was a prime example of a nation whose science aspired so much to be abstract that it failed in practicalities. Mechanics, for example, was much better known in England, and though in France 'many have acquired a profound knowledge of the higher branches of mathematics [,] the more elementary part of mathematics, which serves for every day's use, which leads to the most useful applications, is far less diffused in France than in England'.[31] As William Grove saw it, though, England suffered from exactly the same failing, made worse by an outdated social elitism which hierarchized knowledge so that useful, practical kinds were devalued, associated with manual labour and therefore abandoned to the lower classes: 'Physical, or at all events experimental, science is tabooed; it is written down 'snobbish', and its being so considered has much influence in making it so: the necessity of manipulation is a sad drawback to the gentlemanliness of a pursuit'[32].

Grove did not directly address Moll's charge that monolingualism was one of the chief failings of British science, preferring to take up a different question of language, attacking what he called 'promiscuous coining' of new terms.[33] Barriers formed by the national languages were a lesser problem for the communication of science than the development of dialects within scientific English. Like Whewell arguing against the fragmentation of the scientific disciplines, Grove issued a sarcastic attack on linguistic fragmentation, proposing as a solution a single centralizing authority:

If this work go on, the scientific public must elect a censor whose fiat shall be final; otherwise, as every small philosopher is encouraged or tolerated in framing *ad libitum* a nomenclature of his own, the inevitable result will be, that no man will be able to understand his brother, and a confusion of tongues will ensue, to be likened only to that which occasioned the memorable dispersion at Babel.[34]

A 'confusion of tongues' already existed, as we have seen, in chemistry. The 1850 edition of Charles Daubeny's *Introduction to the Atomic Theory* included a handy table of translation between the chemical notations of Dalton, Berzelius, and the algebraic system, suggesting that in

---

[31] [Moll], *On the Alleged Decline*, p. 12.

[32] [Grove], 'Physical Science in England', p. 516. Grove is using 'snobbish' (a word of very recent coinage) in a sense which was soon to become obsolete: that of a lower-class person or (in Cambridge parlance) a person of the town rather than an academic.

[33] Ibid., pp. 523–4.

[34] Ibid., p. 524. For Whewell on the scientific disciplines, see above, p. 103–12.

chemistry at least, Grove's argument was not too greatly exaggerated. The translations are only possible because Daubeny also includes the plain English names for the compounds and the elements (such as 'water' and 'hydrogen'). Problems of fragmentation, mutual comprehension, class politics, and theory-ladenness were even more extreme where the things to be discussed lacked this kind of commonly accepted name. Baptismal moments in early nineteenth-century science exposed fundamental ideological attitudes and an example of one such moment that involved describing a difficult spatial concept is the focus of the next part of this chapter.

Scientific vocabulary necessarily participates in the cultural milieu of its use; newly-coined scientific terms, like any others, are responsive to the time of their invention. Inventors of new scientific terminology sometimes feel that this historical responsiveness makes their terms vulnerable to attack or rejection from members of the language-user community: the terms simply do not seem old enough to be authoritative. To avoid this, they often seek to root their neologisms in the past, especially in the distant past. In so doing, they necessarily appeal to models of authority derived from non-scientific sources. The next section of this chapter looks in detail at a case of scientific baptism in which Whewell was involved, and in which the inventors of terms sought to mask the newness of their vocabulary with an appeal to the classics. The new terms in question were being developed with layers of spatiality inherent in them: they sought to baptize spatial phenomena, and they asserted claims about the spatial practices of the sciences to which they belonged.

TWO KINDS OF PLATING

During the early 1830s Michael Faraday was investigating the way in which electrical current moved through substances. In a series of sensational experiments he demonstrated that electrical and magnetic forces were related, and that a magnetic field could be made to induce an electric current. He was attempting to show that all known forms of electricity, both current and static, were manifestations of the same basic force.

In the course of his experimentation in 1834 Faraday used the process of electrochemical decomposition, or electrolysis. By passing an electrical current through some mixed solutions, a chemical reaction was produced which could be used to decompose the solution into its constituent parts.[35] This process was to form the basis of a number of important technological applications, including electroplating, by which items made of cheap metal were given a very thin coating of a more valuable material, making—among other things—silverware available to the newly aspirant but not yet wealthy Victorian middle classes, and incidentally giving contemporary satirists an instantly recognizable motif for social snobbery. But in 1834 electrolysis was an experimental rather than an industrial procedure and one which lacked a vocabulary. One of the crucial problems with designing a vocabulary for it was that there was as yet no universally agreed explanation of the nature of electricity. Faraday wanted his work to gain acceptance with the widest possible audience, so could not afford to alienate experimenters who held a different theory of electricity from his own. He therefore intended that the terms he used should be uncontaminated by any particular account of what electricity was.

Faraday's struggle with terminology was, as it almost always is, conceptual. It was also rhetorical, since he was trying to couch his disagreement with contemporary theories of electricity in language which would be nonetheless comprehensible and acceptable to his audience.[36] Though he was aiming for theory-free nomenclature, he did not attempt to expunge metaphor from his terms. On the contrary, the imagery in his terminology was very useful, operating both privately and publicly, as a way of modelling within his own imagination as well as a heuristic device and a tool of persuasion. For all these reasons, the baptism of 'anode' and 'cathode' in 1834 is of real importance in an understanding of nineteenth-century science and its relationships with wider culture.

Changes in accounts of the nature of electricity were one of the pointers towards the new shift away from mechanical, material explanations of the forces that operate in the universe, a shift which in the mid-century would culminate in field theory's radical rethinking of

[35] For a summary of the experiments that led to and surrounded this discovery, see Stanley M. Guralnick, 'The Contexts of Faraday's Electrochemical Laws', *Isis* 70 (1979), 59–75, esp. pp. 60–3.

[36] See for instance L. Pearce Williams, *Michael Faraday: A Biography* (London: Chapman and Hall, 1965), pp. 263–4, for comment on Faraday's ability to advance his views, the content of which was frequently in direct opposition to prevailing scientific belief, through the language of that belief, and his consequent success in establishing the place of his work in debate.

what matter and space are. In some ways electrolysis was a profoundly material discovery: it dealt with the distinguishing and separating of different elements within a compound, somewhat like a scientific version of the myth in which Psyche is asked to sort a roomful of seeds into their different kinds. It was certainly taken up very rapidly by industry. But seen in another light, electrolysis helped Faraday—and with him, eventually, the scientific establishment—to move away from the mechanical model of the nature of electricity as analogous to water flowing along channels and towards an understanding that matter and space are not two different kinds of thing, but different intensities of the same thing. It is not at all surprising that it was difficult to baptize the details of this new discovery, because the names would necessarily have to engage with this as yet unarticulated fundamental shift in thinking about space. It is one thing to use spatial metaphors to discuss abstract concepts such as disciplines and kinds of knowledge, but quite another if they have to negotiate between different basic understandings of space itself. This is one reason why the terminological conversation between Whewell and Faraday in 1834 is of such importance in tracing the history of space in the nineteenth century.

At stake was the highly spatial question of whether electricity really *flowed*. Early theories of electricity had envisaged it as a kind of liquid, or as two different sorts of fluid substance. Faraday by this time had concluded privately that electrical force did not actually 'flow' at all; but the basic terms derived from the fluid hypothesis—including 'current'—were already established in the scientific vocabulary. Faraday had to invent terms that would be conformable to this strand of existing nomenclature but that would not commit him to its underlying theory.

The immediate task was to name two parts of equipment: the electrodes, metals which are attached to the positive and negative poles of the battery and which are submerged in the solution to be electrolyzed. Faraday attempted to generate terms alone and in consultation with friends, but finally wrote in the spring of 1834 to William Whewell, who was well known not only as a man of wide scientific interests, influence, and standing, but also as a philologist who had already contributed to the naming of several recent discoveries—for example, he had coined the terms pliocene, miocene and eocene for use by Lyell in his *Principles of Geology* (1830–3).[37]

---

[37] See Pearce Williams, *Michael Faraday*, p. 261.

Faraday states explicitly that 'I think you will agree with me that I had better not give a new word than give one which is not likely to enter into common use.'[38] Explaining to Whewell that 'it is essential to me to have the power of referring to the two surfaces of a decomposable body by which the current enters into & passes out of it', Faraday continues:

In searching for a reference on which to found these I can think of nothing but the globe as a magnetic body. If we admit the magnetism of the Globe as due to Electric currents running in lines of latitude their course must be according to our present modes of expression from East to West [...]. I think therefore that if I were to call c the *east-ode* & b the *west-ode* I should express these parts by reference to a natural standard which whatever changes take place in our theories or knowledge of Electricity will still have the same relation. But Eastode & Westode or Oriode & Occiode are name[s] which a scholar could not suffer I understand for a moment and *Anatolode* and *dysiode* have been offered me instead.[39]

The basic model for Faraday's conception of electrical current is that of the terrestrial globe. The 'lines of latitude' he mentions are ambiguous: he is referring both to the infinite number of lines which can be drawn on a spherical object orthogonally to any one of its diameters, and also to the fixed number of lines of latitude used in navigation, which are purely imaginary, but seductively concrete in appearance. The globe is a good analogy partly because of the ambiguous, part-material and part-metaphorical nature of the way the earth is coded and decoded by navigational and geometric means. Faraday realized that his terms must be similarly ambiguous, allowing for the possibility of interpretation of their referents as either concrete or conceptual. The terms are required to translate invisibly between opposed theories of the nature of electricity.

But Faraday was anxious not only to free his new terms from limiting dependence on any particular theory of electricity but also to equip them with the philological respectability necessary to induct them into a wider language-user community. This latter desire received Whewell's special approbation:

new terms inevitably arise, and it is very fortunate when those upon whom the introduction of these devolves look forwards as carefully as you do to the general bearing and future prospects of the subject; and it is an additional

---

[38] Faraday to Whewell, 3 May 1834, in *Correspondence of Michael Faraday*, II, p. 181.
[39] Faraday to Whewell, 24 April 1834, in *Correspondence of Michael Faraday*, II, pp. 176, 177. I have accepted James's editorial addition of full stops without marking them with square brackets.

advantage when they humour philologists so far as to avoid gross incongruities of language.[40]

Faraday's prolonged struggles with new nomenclature, while constantly acknowledging the importance of euphony and good classical pedigree, were at first primarily concerned with the avoidance of theory-ladenness, a quality he noted with some disapprobation in other scientists' terms:

> I cannot help thinking it a most unfortunate thing that men who as experiment-alists & philosophers are the most fitted to advance the general cause of science & knowledge should by the promulgation of their own theoretical views under the form of nomenclature notation or scale actually retard its progress—It would not be of so much consequence if it was only theory & hypotheses which they thus treated but they put facts or the current coin of science into the same limited circulation when they describe them in such a way that the initiated only can read them.[41]

This desire for lexical purity, for freedom from intrusion by theory, appears unappeasable to literary criticism: what baptism could possibly avoid introducing any connotative qualities? And Faraday realized that in practice this kind of purity was impossible. Instead, he wanted his terms to acknowledge their own connotativeness. He made an important distinction between two kinds of terms. The first kind were explicitly based on the principle that electricity could be thought of as moving, and expressed this movement in metaphorical ways. The second kind, which he wanted to avoid, lacked the implied caveat that metaphoricity provided. So an element of metaphor was acceptable, provided it could not be mistaken for a real theory. In fact it was all but impossible to avoid terms which used the water metaphor, even in his

---

[40] Whewell to Faraday, 25 April 1834, in *Correspondence of Michael Faraday*, II, 178–9. It was an unconscious compliment to Faraday that Whewell, who had just finished reviewing Mary Somerville's *On the Connexion of the Physical Sciences* for the *Quarterly Review*, recycled two of the key tropes—the bearings and the prospects of science—from the praise he gave Somerville (see above, p. 105–7).

[41] Faraday to Whewell, 21 February 1831, in *Correspondence of Michael Faraday*, I, p. 549. This letter refers to Whewell's paper, 'On the Employment of Notation in Chemistry' (derived from Whewell's pamphlet *An Essay on Mineralogical Classification and Nomenclature; with Tables of the Orders and Species of Minerals* (Cambridge: Smith, 1828)) published in the *Journal of the Royal Institution*, 1 (1830–1), 437–53. In this paper Whewell had argued against the introduction of foreign modes of notation into chemistry without substantial modification to accord with algebraic usages. See also Faraday's explanation on the grounds of theory-ladenness for disliking Ampère's term 'electro-dynamic': Faraday to Whewell, 19 September 1835, in *Correspondence of Michael Faraday*, II, p. 278.

correspondence with Whewell; all that could be done was to highlight
its metaphoricity. In one of the first letters, for instance, Faraday writes:
'*admitting the usual mode of expression & talking of a current* of Electricity
proceeding from the positive pole P through [...] to the negative pole
N [...] '.[42] It is clear from later sections of this same letter and elsewhere
in Faraday's electrical research at this time that he did not in fact accept
the idea of a flowing current as anything more than a metaphor. The
fact that he was prepared to adopt it in order to render his work
assimilable into professional debate is a mark of his rhetorical skill and
his understanding of the linguistic politics of science.

Faraday had tried out several pairs of terms already. Though he
struggled to keep them free of limiting theory, he was happy for them
to make allusions of other kinds. He toyed with the names 'voltode'
and 'galvanode', feeling that they suffered from no inherent dependence
on a particular theory and, presumably, that they honoured two experi-
mentalists whose reputations in electrical research were unquestionable.
But Whewell dismissed them instantly. They were freaks in the course
of the debate.

In his 1840 *Philosophy of the Inductive Sciences*, Whewell gives a brief
and historically misleading account of the rejection of these terms, and
of one of Faraday's other initial ideas, alphode and betode:

A person who did not see the value of our present maxim, that descriptive
terms should be descriptive in their origin, might have proposed words perfectly
arbitrary, as *Alphode* and *Betode*: or, if he wished to pay a tribute of respect to
the discoverers in this department of science, *Galvanode* and *Voltaode*. But such
words would very justly have been rejected by Mr. Faraday, and would hardly
have obtained any general currency among men of science.[43]

In fact these terms were rejected by Faraday because Whewell would
not countenance them. The reasons Whewell ascribed to Faraday were
his own. Scientific terms like 'anode' and 'cathode'—the ones Faraday
eventually settled on—are designed to appear to fit naturally into the
language and to minimize their novelty. They should seem at once
deeply embedded in the past of the language (and of past languages)
and yet at the same time ahistorical—free of the non-linguistic history
which language communicates. The aim of the Whewell–Faraday

---

[42] Faraday to Whewell, 24 April 1834, in *Correspondence of Michael Faraday*, II,
p. 176. Italics added.
[43] William Whewell, *Philosophy of the Inductive Sciences, Founded Upon Their History*,
2 vols (London: Parker, 1840) I, p. xcv.

correspondence was to generate terms which effaced the correspondence (their real history) itself, appearing so rooted in and tied to the language that it would not occur to the reader to wonder about the circumstances of their composition. This is why it is significant that by 1840 Whewell was able to rewrite the history of the composition of 'anode' and 'cathode' in order to suggest that no alternative was ever thinkable. However, those suppressed circumstances of composition nonetheless had a determining role in the creation of terms. A major factor in those circumstances is a set of choices about authority—none of those choices scientific, despite the fact that the clear purpose of the terms is to gain and guarantee scientific authority.

The difficulty over 'anode' and 'cathode' was, at bottom, the difficulty of expressing a pure spatiality. Faraday wanted to indicate movement through space without implying that any particular substance, or any substance at all, was actually moving. The terms must allow for the possibility that the space, and the movement of electricity through it, were conceptual rather than physical. So a great many of the words Whewell offered Faraday are directional: anode and cathode signify way up and way down (or rising way and setting way); orthode and anthode, direct way and opposite way; dexiode and sceode, right way and left way.[44] The underlying model suggested by all these terms is vectoral rather than scalar—that is, one involving not just the idea of movement but of movement in a particular direction. But Whewell was not suggesting that current really flowed up and down, or left and right: the directions were there to serve as paired opposites rather than as indicators of events in the real world. As he put it in *Philosophy of the Inductive Sciences*, 'the processes have not reference to any opposed points, but to two opposite directions of a path'.[45] This is an interesting and important kind of spatiality: a non-teleological space in which one can discern opposite directions but not any destinations to which those directions point. One of Whewell's criticisms of the alphode/betode and voltode/galvanode suggestions was that they did not suggest this kind of space effectively:

it is very desirable in this case to express an *opposition*, a contrariety, as well as a difference. The terms you suggest are objectionable in not doing this. They are also objectionable it appears to me, in putting forwards too ostentatiously the

---

[44] Ibid., I, p. xcv; Whewell to Faraday, 25 April 1834, in *Correspondence of Michael Faraday*, II, p. 179.
[45] Whewell, *Philosophy of the Inductive Sciences*, I, p. xcv.

arbitrary nature of the difference. To talk of Alphode and Betode could give some persons the idea that you thought it absurd to pursue the philosophy of the difference of the two results [...].[46]

He suggests that inadequately metaphorical terms would be both theory-laden (stressing the arbitrariness of the distinction between the two surfaces) and inhibitive of further scientific investigation. It was important to maintain metaphoricity.

The Faraday biographer L. Pearce Williams gives an analysis of the Whewell–Faraday correspondence which usefully elucidates the different models and theories Faraday adopted during the course of his researches. Pearce Williams comments on Faraday's early suggested terms, eisode and exode (way in and way out), that

> What Faraday really meant to convey was direction, not a process. By defining the electric current as '*an axis of power having contrary forces, exactly equal in amount, in contrary directions*' [...], he had wanted to indicate merely a general polarity, extending throughout the electrolyte, of which the poles were the termini. 'Into' and 'out of', however, implied that something material was passing through the electrolyte, and this was precisely what Faraday wished to avoid.[47]

Faraday's terminology is intended to produce a purely spatial meaning. He wants to indicate direction without basing that direction on any given dimension, and without even implying that any entity or substance is in fact moving. In effect, Faraday's understanding of electricity demanded a new understanding of the difference between space and matter: the baptismal problem was finding a way of expressing a relationship existing in a space so pure and abstract that there could be no question of associating it with movement or matter.

Another of Faraday's basic terminological aims was almost certainly to acquire for his ideas the academic authority of the classics. He mobilized Whewell's classical scholarship to grant a sort of linguistic birthright to his innovations. Even Whewell's name was enough to silence initial

---

[46] Whewell to Faraday, 6 May 1834, in *Correspondence of Michael Faraday*, II, p. 184.

[47] Pearce Williams, *Michael Faraday*, p. 261. The italicized words are Faraday's, and are quoted from Michael Faraday, *Experimental Researches in Electricity*, 3 vols (London: Taylor & Francis, 1839–55), I, series V, ¶ 517, p. 148. Pearce Williams' description of Faraday's formulation 'an axis of power having contrary forces, exactly equal in amount, in contrary directions' as a definition is perhaps overstating Faraday's claim: Faraday himself only argued that 'the influence present in what we call the electric current, [...] may perhaps best be conceived of as' the formulation quoted.

opposition to Faraday's choice of nomenclature: as he wrote to Whewell in May 1834:

I had some hot objections made to [the terms] here and found myself very much in the condition of the man with his son and Ass who tried to please every body; but when I held up the shield of your authority it was wonderful to see how the tone of objection melted away […] [48]

Gillian Beer has argued that the nineteenth-century emphasis on classics in boys' education, and its almost entire absence from the education of girls, gave Greek and Latin the status of exclusionary languages, used by a masculine, middle- and upper-class group. Consequently,

the practice of Victorian scientists of citing classical writing in their work serves several functions: some social, some illustrative, some argumentative. Such allusion effortlessly claimed gender and class community with a selected band of readers; it implied a benign continuity for scientific enquiries with the imaginative past of human society; it could figure the tension between objectivity and affect. [49]

Alluding to a classical language served two contrary but related purposes. As Beer indicates, the effect of classical allusion was to reinforce an exclusion of (for instance) female, or working-class, readers. It might seem peculiar and perhaps downright unpleasant of Faraday, an artisan autodidact with no classical education, to resort to ancient Greek with these exclusionary effects. But as Laura Otis points out, 'those scientists who did not come from the socially privileged classes had even more to gain by establishing reputations as men of humane learning'.[50] Faraday never tried to pass himself off as knowing more of the classical languages than he really did; in 1831, having invented the Greek-derived term 'electrotonic state' for the condition of matter involved in electrical conduction, he seemed surprised at his own linguistic confidence: 'am I not a bold man, ignorant as I am, to coin words' he asked, then added 'but I have consulted the scholars'.[51] It was important for the acceptance of his results that his linguistic innovations be as unexceptional as possible. And seen in another light, allusions to Latin and Greek

---

[48] Faraday to Whewell, 15 May 1834 (*Correspondence of Michael Faraday*, II, p. 186).
[49] Beer, 'Translation or Transformation? The Relation of Literature and Science', pp. 175–6.
[50] Laura Otis, 'Introduction' to *Literature and Science in the Nineteenth Century: An Anthology* (Oxford: Oxford University Press, 2002), pp. xvii–xxviii (p. xix).
[51] Faraday to Richard Phillips, 29 November 1831, in *Correspondence of Michael Faraday*, I, p. 590. Faraday signs off 'Excuse this egotistical letter […] ' (p. 591).

appeared to be *in*clusive. By basing one's language on Greek or Latin one could claim to be transcending the nationalism of vernacular languages and rooting one's practice instead in the shared language of learned communities. Whewell, as we have seen, promoted this use of Latin. Indeed he advocated the inclusionary ideal of the classics in his account of scientific language in *The Philosophy of the Inductive Sciences*. Terms derived from Greek and Latin

are readily understood over the whole lettered world. [...] The advantage of such terms is, as has already been intimated, that they constitute an universal language, by means of which cultivated persons in every country may convey to each other their ideas without the need of translation.[52]

All the terms Faraday considered for naming the apparatus of electrolysis were developed from classical roots, using the Greek root ὁδός (a way) for their terminations. In their tacit insistence on claiming a linguistic classical inheritance for Faraday's innovations, Whewell and Faraday were applying on behalf of the new ideas for membership of the gender and class community described by Beer. They were also associating the recent discoveries of science with the authority of the ancient past. Whewell and Faraday sought to provide the terms with an unassailable paternity, to weave them into an authoritative, traditional discourse of knowledge and privilege.

The value of Whewell's coinages was not universally accepted. William Grove attacked Whewell explicitly in his essay in *Blackwood's* in 1843 for his work as a baptizer of scientific ideas, citing as examples of his bad practice one word which, in the event, went on to be universally adopted—'physicists'—and one which did not—the rather bizarre 'idiopts', meaning colour-blind people. Grove felt that although Whewell's 'scientific position and dialectic turn of mind may fairly qualify him to be a word-maker', he seemed 'peculiarly deficient in ear'. 'Idiopts', in particular, would 'scarcely thank the Master of Trinity College for approximating them in name to a more numerous and more unfortunate class'. The sibilants in 'physicists' 'fizz like a squib'. 'Euphony [was] wantonly disregarded', and in setting such a poor example, Whewell was indirectly responsible for the coinages of 'authors of smaller calibre', in which 'classical associations are curiously violated'. Grove's example here was 'platinode', meaning the negative pole of a voltaic cell, which he pointed out was 'Spanish-American joined to ancient Greek'.[53]

Whewell would have insisted that his own coinages did not violate classical associations: 'I may mention too that *anodos* and *cathodos* are good genuine Greek words, and not compounds coined for the purpose.'[54] 'Genuine' and 'coined' are two keywords in the analogy between language and money which, since Locke, had been very influential in British philosophy of language. Grove used this analogy to discuss the institutional politics of science, linking the privilege of having one's coinages accepted to recognized scientific prestige: 'As coinage of money is the undoubted prerogative of the crown; so generally coinage of words has been the undoubted prerogative of the kings of science—those to whom mankind have bent as to unquestionable authority.' In the present day, though, 'every man has his own mint, and although their several coins do not pass current very generally, yet they are taken here and there by a few disciples, and throw some standard money out of the market'.[55] But he does not explore the question of what defines 'standard money' and by what guarantees it is backed. Some currencies simply go out of date. Whewell, who naturally wanted his coinages to be durable, had a great deal invested in the argument that terminology can somehow transcend the theory-systems which support it. Terms which 'express relations really ascertained to be true' cannot 'lose their value by any change of the received theory':

They are like coins of pure metal, which, even when carried into a country which does not recognize the sovereign whose impress they bear, are still gladly received, and may, by the addition of an explanatory mark, continue part of the common currency of the country.[56]

The widespread deployment of terms drawn from the semantic field of money added a mercantilist concern with provenance and authenticity to the linguistic quest for translatability, and complicated the debate about what level of metaphoricity was acceptable in scientific terminology. These terms try to suggest that, in the same way that the monarch's head on a coin is irrelevant provided the coin is agreed to be good, different systems for guaranteeing the validity of knowledge are

---

53 [William Grove], 'Physical Science in England', p. 524.
54 Whewell to Faraday, 25 April 1834, (*Correspondence of Michael Faraday*, II, p. 179).
55 [Grove], 'Physical Science in England', p. 524.
56 Whewell, *Philosophy of the Inductive Sciences*, I, p. lxvi.

irrelevant provided the basic nature of the knowledge itself is universally recognized. This argument severely understates the importance of precisely those guaranteeing systems in the production and promulgation of agreement about what the nature of the knowledge in fact is, but it has always been an important strand in scientific rhetoric to minimize the history and sociology of the production of scientific knowledge in favour of claims to universality and elementarism.

Some years later in a letter to Faraday, Whewell played on the connection of the language-as-coin metaphor and the importance of scientific language in instituting what might now be called a paradigm shift:

Such a coinage has always taken place at the great epochs of discovery; like the medals that are struck at the beginning of a new reign:—or rather like the change of currency produced by the accession of a new sovereign [...]. [57]

But despite this emphasis on the fresh start offered by new names, the Faraday–Whewell correspondence demonstrates that one of the most important of all the rhetorical purposes served by using classical languages in the formation of new terminology was the erasure of the historical moment of the baptism: ancient languages hide the modernity of the coining. Durability depended in part on this erasure.

As Whewell's inventions 'anode' and 'cathode' made their way through European science to become accepted terms, electrolysis was being developed into electroplating and being adapted by an enormously wide range of Victorian industries. And in literary writing, electroplating was rapidly being adopted as a symbol of the shoddiness and superficiality of the age. Electroplated goods were used as an emblem of insincerity, bad enough where they simply pretended to wealth, but especially disgraceful on occasions where deep emotion or stately ceremonial was called for. When the eponymous hero of Lady Stuart-Wortley's 1844 'comedietta' *Ernest Mountjoy* orders a coffin, he tells the bewildered tailor to whom he is mistakenly giving the commission:

I wish it to be thickly covered with polished brass nails, and also it must have large ornamented handles; let it be adorned with silver-plated cherubs' heads,

[57] Whewell to Faraday, 14 October 1837, *Correspondence of Michael Faraday*, II, p. 464.

and lined splendidly with full draperies of the richest white satin through-out, and, possibly, an outward surtout of *lead* might be desirable and advan-tageous.[58]

These silver-plated cherubs' heads must have been a feature of funerals in 1844. Dickens's undertaker Mr Mould in *Martin Chuzzlewit* is gratified in the same year by a lavish order

to provide silver-plated handles of the very best description, ornamented with angels' heads from the most expensive dies. To be perfectly profuse in feathers. In short, sir, to turn out something absolutely gorgeous.[59]

Electroplating made 'something absolutely gorgeous' in silver available to the upwardly mobile, even if that something was a coffin-handle.

Although it began as a sign of phoniness or insincerity, within a generation electroplating started to lose its low moral status. By 1878 Charles L. Eastlake's manual on interior design, *Hints on Household Taste*, was holding electroplated goods up as a standard of respectab-ility by which other domestic technologies of transfiguration, such as veneering, should be judged:

Veneering has been condemned by some writers on the same grounds on which false jewellery should (of course) be condemned. But the two cases are scarcely analogous. If we are to tolerate the marble lining of a brick wall and the practice of silver-plating goods of baser metal—now too universally recognised to be considered in the light of a deception—I do not see exactly how veneering is to be rejected on 'moral' grounds.[60]

Eastlake's readers might not consider electroplated household goods 'in the light of a deception', but other people were electroplating other objects with definitely criminal intention. Counterfeiters were able to electroplate their worthless coins, covering them with a very thin layer of silver that gave the unwary the impression they were silver all the way through.[61] This adaptation of the technology gives us an alternative

[58] Lady Stuart-Wortley, *Ernest Mountjoy, A Comedietta* (London: n. publ., 1844), p. 33.
[59] Charles Dickens, *The Life and Adventures of Martin Chuzzlewit* (1844), ed. Patricia Ingham (London: Penguin, 1999), p. 305 (ch. 19).
[60] Charles L. Eastlake, *Hints on Household Taste: The Classic Handbook of Victorian Interior Design*, 4th rev. edn (London: Longman, Green, 1878; repr. Mineola, N.Y.: Dover, 1986), pp. 56–7.
[61] The process is described in Iwan Rhys Morus, *Frankenstein's Children: Electri-city, Exhibition, and Experiment in Early Nineteenth-Century London* (Princeton, N. J.: Princeton University Press, 1998), p. 176.

way of reading the discourse of money that pervades Whewell's, Grove's and other contemporary writers' analysis of scientific baptisms. Faraday wanted his terms to pass as tender in all the major theories of electricity. He wanted them to be, as it were, hollow, not carrying the weight of any particular theory. Whewell spread a very thin layer of classical authenticity over the top of these hollow coins and made them—unlike the products of the counterfeiters—all but irreproachable. The metaphor of the electrical current was transmogrified into the reality of currency.

PART II

THINKING ABOUT SPACE

5

Aspiring to the Abstract: Pure Space and Geometry

What did 'space' mean for the early nineteenth century, and how did scientific accounts of space interact with other cultural ideas about it? One way to begin to develop an answer to the first question would be via a history of philosophy. This kind of answer might centre on contemporary support for and opposition to Kant's view that space is a fundamental category of human knowledge rather than a substance or other thing having real existence. The impact of Kant's arguments on European science in this period has certainly been the subject of a number of illuminating recent studies, but they do not help us with the question of what Kantian space meant in a wider cultural context, not with what other ways of thinking about space were available.[1] To trace the interactions of scientific and other cultural ideas about space requires a much broader methodology, one which rests on a model of dissemination like those developed for work on literature and the earth and life sciences in the later Victorian period.

One of Gillian Beer's great contributions to the development of a cultural history of Victorian science has been her demonstration that such investigations need to be based not on a traditional 'history of ideas' model of dissemination but on one that recognizes the diffuseness and variety of means of reception of ideas through the culture(s) of nineteenth-century Britain.[2] Probably the term 'dissemination' itself is

[1] Recent helpful philosophical studies of Kant's impact on the nineteenth century physical sciences include Eric Watkins's collection *Kant and the Sciences* (Oxford: Oxford University Press, 2001) and Michael Friedman's *Kant and the Exact Sciences* (Cambridge: Harvard University Press, 1992).

[2] For example, in her essay 'Translation or Transformation? The Relations of Literature and Science', in *Open Fields: Science in Cultural Encounter* (Oxford: Clarendon Press, 1996), pp. 173–95, Beer sets out a number of central methodological questions which I take to be shared by most scholars in nineteenth-century literature and science

inappropriate for this purpose, too apt to suggest a planned or intended transmission of ideas. A more accurate sense of the movement of ideas from context to context within the period would emphasize the accidental, the partial, and the metaphorical. In 1855, at the end of the period this book focuses on, William Whewell complained passionately:

there are at present a number of scientific words current amongst us, which are applied with the most fantastical and wanton vagueness of meaning, or of no meaning. [...] There are words [which] every one pursuing fancies which are utterly out of the sphere of science, seems to think he may use just as he pleases. *Magnetism* and *Electricity*, and the terms which belong to these sciences, are especially taken possession of for such purposes, and applied in cases in which we know that the sciences from which the names are '*conveyed*' have not the smallest application. Is Animal Magnetism anything? Let those answer who think they can: but *we* know that it is not *Magnetism*.[3]

Whewell was speaking to a prestigious gathering at the Royal Institution, in the presence of Prince Albert. Even so august and powerful a group was, he noted, quite unable to control the travel of scientific words and ideas through the culture, and unable to prevent their mutation as they travelled. For Whewell the absence of this kind of control was lamentable, but we can identify it as characteristic of the vitality of both scientific and other cultures within nineteenth-century Britain. Whewell's outburst bears witness to what we must recognize as the uncontrollable nature of the processes of cultural borrowing, appropriating, half-digesting, and half-comprehending, processes that respect no boundaries of disciplinary dignity, vocabulary, or ownership and turn literature into science, science into literature, and all into the fertile culture of a society with widespread literacy and access to publishing.

The second part of this book aims not to misrepresent this uncontrollableness. Instead of tracing the flow of (say) Kantian theories of

and which tease out the subtle historiographical work demanded by attention to cultural dissemination rather than a search for the origins and genealogies of ideas: 'How to articulate the interactions between apparently remote preoccupations? How to analyse the close written relations between authors who probably never read each other's work? How to explain the concurrent appearance of similar ideas in science and in literature without inevitably forging causal links? And how to avoid stabilizing the argument so that one form of knowledge becomes again the origin of all others?' (p. 177).

[3] William Whewell, 'On the Influence of the History of Science upon Intellectual Education', in *Lectures on Education: Delivered at the Royal Institution of Great Britain* (London: Parker, 1855), pp. 3–36 (pp. 33–4).

space through *Naturphilosophie*-influenced British science, the following chapters try to show some of the conflicting slew of meanings that 'space' had for the educated general reader, focusing particularly on space in its extreme forms. In this first chapter the central topic is geometry and the abstract space to which it claimed to give access. Then Chapter 6 examines the pure space of emptiness, and looks at ways in which void space acquired new potencies in the physical sciences in the early nineteenth century. The last chapter returns to thinking about the kind of territorial spatialities discussed in the first part of the book, but this time focusing on space and *dis*order rather than order, looking at representations of chaos in contemporary writing.

Foucault sees the late eighteenth century as 'the moment when a considered politics of space was starting to develop': at that moment 'the new achievements in theoretical and experimental physics dislodged philosophy from its ancient right to speak of the world, the cosmos, finite or infinite space'.[4] By the period this book focuses on, a generation or so had passed since this decisive moment, and yet at the same time as Foucault's 'considered politics of space' was establishing itself, a counter-space was in evidence, one that attempted to combine the authority of physics and philosophy. This kind of space was aligned with the apolitical, the transcendent, and the immaterial. It was in many ways a product of European Romanticism, and in particular was one of the effects of Kant's legacy in aesthetics and in science.

It would be entirely possible to give an account of early nineteenth-century conceptions of space from a strictly materialist point of view.[5] This was a period in which major changes occurred in British culture's concepts of lived, geographical space. The impetus to sustain and expand a global empire; at home, massively increasing industrialization and urbanization, driving domestic populations into new geographical formations; and developments in technology which transformed the realities of travel and architecture: all these were factors profoundly affecting early nineteenth-century Britain's relationship with material space. And many of these newly produced spaces and spatial practices have received attention from historians and literary critics trying to

---

[4] Foucault, 'The Eye of Power', in *Power/Knowledge: Selected Interviews and Other Writings, 1972–1977*, ed. Colin Gordon, trans. Colin Gordon *et al.* (Brighton: Harvester, 1980), pp. 146–65, p. 149.

[5] Here I am using 'materialist' in its Marxist and post-Marxist political sense rather than its early nineteenth-century anti-religious sense.

describe the effect on Victorian culture and, particularly, Victorian identity formations, of changes in the meaning of 'public' and 'private', 'domestic', 'foreign', and other spatial concepts.

But alongside these new spatial practices, so deeply reflective of new material conditions, were the cultural manifestations of a quite different kind of spatial imagination, one apparently indifferent to these radical shifts in lived geography. This was an immaterial spatiality, a network of imaginative formations clustered round the concept of ideal space, a kind of space that the early nineteenth century often called 'abstract'. Abstract space defies direct expression in material forms but can be alluded to in writing, which maintains an uneasy relationship with the sensory and the imaginary. In Chapter 3 we saw William Whewell and others lamenting the fragmentation of the arts and sciences into a multitude of mutually incomprehensible disciplines and sub-disciplines, and urging instead the dismantling of boundaries to re-form a single, unified space, often metaphorized as a 'field'. 'Abstract' space is in a sense the ideal from which this aspiration towards unity is derived. Abstract space was a frequently invoked object of the cultural imagination during the first decades of the nineteenth century, providing consoling and high-minded images of a boundaryless realm free from human passions but accessible to the human mind. This chapter is about that immaterial space: how the attempt to write about it engendered particular kinds of metaphor, how it engaged or refused to engage in political and other public debate, and why it was so widely attractive to writers in such different disciplines and genres.

It is very difficult to engage with this nineteenth-century ideal abstract space using twentieth-century critical tools which so frequently emphasize the material at the expense of the immaterial. The dominant kind of critical theory of space in the late twentieth- and early twenty-first century emphasizes a materialist approach in order to express its political concerns. Henri Lefebvre's 1974 book *La production de l'espace*, translated in 1991 by Donald Nicholson-Smith, is among the materialist texts which have most powerfully influenced recent work on spatiality. As a Marxist, Lefebvre argues that space is produced by societies in a way somewhat analogous to that in which commodities are produced, and that it is wrong to imagine space as existing prior to the society producing it. His emphasis on the need to engage with the specifics of the practices that produce space in a given society leads him to take issue with what he sees as theory's unrigorous facility for conflating social (historical, geographical) space with mental or otherwise non-material

space. Responding sceptically to Foucault and Derrida, Lefebvre warns
that these theorists have abandoned attention to material space in favour
of a less specific, less political, and ultimately less useful idea of 'space'
as something existing in the mind:

Most if not all authors ensconce themselves comfortably enough within the
terms of mental (and therefore neo-Kantian or neo-Cartesian) space, thereby
demonstrating that 'theoretical practice' is already nothing more than the
egocentric thinking of specialized Western intellectuals—and indeed may soon
be nothing more than an entirely separated, schizoid consciousness.[6]

Lefebvre sees the interests of the dominant class at work in this cosy
concentration on the immaterial. Derrida, Barthes, and Kristeva are part
of a school which, he argues, is

forever promoting the basic sophistry whereby the philosophico-epistemological
notion of space is fetishized and the mental realm comes to envelop the social
and physical ones. Although a few of these authors suspect the existence of,
or the need of, some mediation, most of them spring without the slightest
hesitation from mental to social.[7]

By privileging mental or conceptual space, these philosophers have
unwittingly served the political aims of the hegemony:

What is happening here is that a powerful ideological tendency, one much
attached to its own would-be scientific credentials, is expressing, in an admirably
unconscious manner, those dominant ideas which are perforce the ideas of the
dominant class.[8]

Instead of this attention to conceptual space, Lefebvre insists on the
need to study the role of daily material practice in the construction of
space.

In Lefebvre's wake, writers on space and critical theory have tended
to endorse his emphasis on the role of habitual social and economic
practices in the production of space. In literary studies, this emphasis
on the material and the social in thinking about issues of spatiality
has influenced the development of a methodology producing studies
of particular spaces, which could perhaps more properly be termed
particular *places*, usually as they existed or were represented at particular historical moments. Victorian London is one of these places;

---

[6] Henri Lefebvre, *The Production of Space*, trans. Donald Nicholson-Smith (Oxford:
Basil Blackwell, 1991), p. 24.
[7] Ibid., pp. 5–6.   [8] Ibid., p. 6.

other studies in Victorian space tend to cluster around domestic areas.[9] These studies have been immensely illuminating about the meaning of representations of particular locations, but because of their commitment to exploring spatial practice—how locations are constructed by daily human activity—they are much less interested in identifying Victorian concepts of space as a thing distinct from the people who inhabit or visit it. Almost no literary critical attention has been paid to nineteenth-century concepts of space other than in conjunction with identity politics or geographical location. The result is that we do not yet know enough about what 'space' signified to the nineteenth century.

Spatial metaphors that draw parallels between ways of living in space and certain kinds of intellectual activity (such as those discussed in previous chapters—metaphors of ownership, invasion, and fragmentation, for example) demand that critical attention be given to the kinds of spatial practice that form the root of the metaphor. Accordingly, a full account of the role of material space in Victorian culture would include histories of such varied topics as changes in property law, methods of military occupation of hostile territory, and developments in urban planning.

But other nineteenth-century ways of writing about space sought to remove the spatial from the realm of the lived, to conceive of space as form rather than practice. This contradicts the Lefebvrean model and that of much recent criticism, in which practice has usually been associated with the material and physical and has hence come to seem more politically and historically expressive than form. Naturally, the notion of a space entirely free of practice may be put to serve in all

---

[9] See for example Thad Logan, *The Victorian Parlour* (Cambridge: CUP, 2001). Several recent critical studies of the spaces of Victorian London have been interested in geographies of deviance, from Simon Joyce's *Capital Offenses: Geographies of Class and Crime in Victorian London* (Charlottesville: University of Virginia Press, 2003) to Judith R. Walkowitz's *City of Dreadful Delight: Narratives of Sexual Danger in Late-Victorian London* (London: Virago, 1992). Barbara Epstein Nord's *Walking the Victorian Streets: Women, Representation, and the City* (Ithaca, N.Y.: Cornell University Press, 1995) and Erika Diane Rappaport's *Shopping for Pleasure: Women in the Making of London's West End* (Princeton, N.J.: Princeton University Press, 2000), among others, focus on the geographies of the female pedestrian in the capital. Lynda Nead's *Victorian Babylon: People, Streets, and Images in Nineteenth-Century London* (New Haven, Ct.: Yale University Press, 2000), another influential study of the spaces of Victorian London, generally takes a similarly street-level approach but makes very interesting diversions beneath and above the street, via sewers and hot air balloons.

manner of practical ways and in all kinds of political contexts, and certainly needs to be understood at least partially within those contexts. But before that can be attempted, we need to try to understand the early nineteenth-century's yearning for space *beyond* the human, beyond the physical and material, and the first step is to learn to recognize it when we see it: and here twentieth-century criticism is often unhelpful.

There is no equivalent in studies of nineteenth-century culture of Simon Varey's splendid *Space and the Eighteenth-Century Novel,* though Franco Moretti's wonderfully stimulating *Atlas of the European Novel* both begins such a project and goes far beyond Varey's bounds.[10] On the whole, where accounts of the geography of Victorian literature, and geography *in* Victorian literature, have addressed spatiality as distinct from particular places, they have done it in relation to subjectivity and selfhood. Psychoanalytic methodologies have been popular for this kind of approach, as in many of the studies of space in Gothic literature.[11] Or historically-based work may adopt a Foucauldian approach, as with Lynda Nead's excellent, detailed study of class and gender formations in nineteenth-century representations of one particular London street. Nead's argument that 'social space [...] is not a passive backdrop to the formation of identity, but is part of an active ordering and organizing of the social and cultural relations of the city' would be echoed by most contemporary scholars writing on space, culture, and identity.[12]

But a focus on identity generally involves—as in Nead's article—a focus on the particular places in relation to which that identity is constructed and experienced. Robert L. Patten, who acknowledges his debt to Franco Moretti, asks the very stimulating question 'At different times during the century, do different kinds of spaces seem to predominate in literary representations?'. He sets up three categories for the kinds of space he thinks will help answer this question: spaces represented in the text ('houses, parks, streets'), spaces that constitute

---

[10] Simon Varey, *Space and the Eighteenth-Century English Novel* (Cambridge: CUP, 1990); Franco Moretti, *Atlas of the European Novel, 1800–1900* (London: Verso, 1998).

[11] See for example Kate Ferguson Ellis, *The Contested Castle: Gothic Novels and the Subversion of Domestic Ideology* (Urbana: University of Illinois Press, 1989) and Pamela J. Shelden, 'Jamesian Gothicism: The Haunted Castle of the Mind', *Studies in the Literary Imagination,* 7 (1974), 121–34.

[12] Lynda Nead, 'Mapping the Self: Gender, Space and Modernity in Mid-Victorian London', in Roy Porter, ed., *Rewriting the Self: Histories from the Renaissance to the Present* (London: Routledge, 1997), pp. 167–85 (p. 167).

the text ('formats'), and 'the spaces, times, and social constructions of readerly activity'. But like almost all literary critical investigations of nineteenth-century space, this approach emphasizes geographical places to the exclusion of 'space' itself.[13]

Certain key spatial concepts (the public/private binary has been particularly productive) are used to organize, classify, or understand key places of the Victorian novel, but explorations of the individual (whether in Victorian literature or in twentieth-century criticism) tend to generate explorations of individual places, and to inhibit thinking about space as distinct from place. Henri Talon's account of Pip's emotional landscape in *Great Expectations* dates from the 1970s but is not untypical of much current literary criticism dealing with Victorian spatiality in conflating space and place. Talon notes that 'space associated with his childhood lives in [Pip's] memory', but goes on to show that it is particular places, rather than space more generally, that matters to Pip: the marshes, Satis House, the forge, the ruined garden. Indeed, Talon quotes G. van der Leeuw arguing that autobiographical space is always non-conceptual: 'When one says space and time, one means one's own life. That is, one means not empty surface, not a chronometer, but myth.'[14] Does this polarization, with lived space at one extreme and abstract space at the other, mean that abstract space is alien or antithetical to selfhood, to the speaking subject? 'Empty surface' is in some ways a good description of the kind of space the early nineteenth century meant by 'abstract' space, which is 'empty' in the sense that none of the things that exist within it can affect or alter it. If van de Leeuw is right and the autobiographic impulse cannot accommodate this kind of space, identity politics and the kinds of criticism that depend on them will surely be unproductive ways of investigating it.

Other kinds of twentieth-century critical theory may provide tools more sympathetic to the project of identifying this abstract space. The theorists Lefebvre attacks for focusing on immaterial space rather than the space of human practice are an obvious starting point. Perhaps the closest analogy in contemporary theory to the early nineteenth-century ideal of immaterial space is the set of various postmodern and

[13] Robert L. Patten, 'From House to Square to Street: Narrative Traversals', in Helena Michie and Ronald L. Thomas, eds., *Nineteenth-century Geographies: The Transformation of Space from the Victorian Age to the American Century* (New Brunswick, N.J.: Rutgers University Press, 2003), pp. 91–206, (pp. 191, 192).

[14] Henri Talon, 'Space, Time, and Memory in *Great Expectations*', *Dickens Studies Annual*, 3 (1974), 122–33 (pp. 125–6).

poststructuralist reinterpretations of Plato's *khora*. In Plato's dialogue *Timaeus*, Critias describes

a third nature, which is space [*khora*], and is eternal, and admits not of destruction, and provides a home for all created things, and is perceived without the help of sense, by a kind of spurious reason, and is hardly matter of belief [...].[15]

Though Derrida and Kristeva, among others, put *khora* to different uses, it remains in their work still identifiably spatial and ideal, and strongly contrasted to the system of distinctions, boundaries, binaries, and hierarchies that constitutes normal epistemological experience.[16] Kristeva especially uses the 'thirdness' or outsideness of *khora* (or *chora* in her transliteration) to envisage this space as a kind of ecstatic maternal realm, an unformed space which pre-exists the symbolic order.[17] Though the postmodern theologian Graham Ward goes so far as to identify cyberspace as 'the realization' of *khora*, more usually *khora* features in contemporary critical theory as a space which is utopian and impossible, unrealizable yet potent.[18] Even in Kristeva's linking of *chora* to the female body, this kind of space is non-physical and all but indescribable; in almost all its guises in contemporary critical theory *khora* is abstract and immaterial. It is at least arguable that this shared idea of an abstract politically or psychoanalytically powerful space is in some ways (perhaps rather attenuated ways) a legacy of nineteenth-century spatial theorizing, especially of the Victorian emotional investment in a non-human immaterial space in which the historical, contingent, and personal are irrelevant.

---

[15] Plato, *Timaeus*, trans. Benjamin Jowett, in *The Dialogues of Plato [...]*, 4 vols (Oxford: Clarendon Press, 1871), II, pp. 513–86 (p. 545). Jowett's introduction to his own translation of *Timaeus* opens with the daunting claim that though of Plato's writing it is 'the most obscure and repulsive to the modern reader', it nonetheless had 'the greatest influence over the ancient and mediaeval world' (p. 467). Jowett, while among the most influential, may not perhaps be the most representative of Victorian Hellenists; but his view of *Timaeus* has had surprising longevity. In the very different context of his postmodern philosophical book on *Timaeus*, John Sallis similarly calls it 'a dialogue of strangeness [and] utterly singular' and acknowledges that though 'one cannot but approach the *Timaeus* with a certain reticence', 'of all the dialogues it is the one that has been most continuously and directly effective' (*Chorology: On Beginning in Plato's Timaeus* (Bloomington: Indiana University Press, 1999), pp. 2–3).

[16] Jacques Derrida, 'Khora', trans. Ian McLeod, in *On the Name*, ed. Thomas Dutoit (Stanford, Calif.: Stanford University Press, 1995).

[17] Julia Kristeva, *Revolution in Poetic Language*, trans. Margaret Waller (New York: Guildford, 1984), pp. 24–5.

[18] Graham Ward, *The Postmodern God: A Theological Reader* (Oxford: Blackwell, 1997), p. xvi.

Instead of attempting a search for the origins of twentieth-century *khoras* in Victorian spatial theory, though, most of this chapter will centre on aspects of geometry, that highly privileged Victorian discourse, and especially its role in early nineteenth-century writing about science. I shall leave it to the following chapter to explore in more detail what might be called the 'mechanics' of abstract space, that is, physical investigations of its properties and metaphysical debates about its nature. But in order that my arguments about the place and prestige of geometry should make sense, it is necessary to outline one strand in the contemporary understanding of abstract space, which will go some way to explaining the reasons for the intense emotional investment in the moral as well as scientific value of geometry among many contemporary commentators.

There is a kind of contradiction in early nineteenth-century writing about abstract space. On the one hand it is described in transcendent terms as taking the mind out of the temporal realm of human passions and leading it to meditate on timeless, perfectly rational certainties. But on the other hand, those influential writers on science who were influenced by German Romantic philosophy and particularly by Kant, understood abstract space as something that existed in the human mind rather than in the world of the senses and of objective phenomena. Putting these two attitudes together, we can see that there is a sort of optimism in the belief that nature is capable of the perfection of abstraction. Of course it is entirely possible to read this reification of the stillness, timelessness, and inaccessibility of abstract space as a political strategy designed to create further mystique around knowledge and thus to retard the process of the democratization of knowledge. Certainly, as we shall see later, some writers argued that geometry (the best-known means of thinking about abstract space) was being used in this way. But there is also a sense in which an opposite political reading is possible: that is, if abstract space has its origins in the mind itself, then it is fundamentally the property of every human being and potentially accessible to anyone's thought processes. Again, we shall see that much was made, rhetorically, of the belief that the study of abstract space, like that of all the abstractions on which science was thought to be based, required no resources beyond reason and concentration, and did not depend on observation of anything in the outside world.

William Whewell addressed the question of the nature of space and human perception of it in his *Philosophy of the Inductive Sciences*. This book is characteristic of early nineteenth-century English idealism in its debt to German Romantic philosophy, especially to Kant: as a major study of Whewell notes, Kant 'is reproduced almost verbatim on space and time in the first 1840 edition of the *Philosophy*'.[19] Whewell and Kant's other English followers took up his argument that the concept of space is not a result of experience, but exists prior to it. Thus, as Whewell explained it, space does not exist 'as an object or thing', but 'as a real and necessary condition of all objects perceived'.[20] It is not because of our sensory experience that we have the category of 'space', but because of a quality inherent within human perception:

we may thus consider space as a *form* which the materials given by experience necessarily assume in the mind; as an arrangement derived from the perceiving mind, and not from the sensations alone. [...] we cannot perceive objects as in space, without an operation of the mind as well as of the senses—without active as well as passive faculties.[21]

Whewell then becomes impatient with the difficulty of expressing these concepts, and attempts to nail the matter with a forceful formula: 'whether we call the conception of space a condition of perception, a form of perception, or an idea, or by any other term, it is something originally inherent in the mind perceiving, and not in the objects perceived'.[22] Not surprisingly, since both were Kantians, Coleridge thought about space in broadly the same way as Whewell. In a letter of 1826, Coleridge constructs a mental experiment intended to show that space 'belongs to the mind itself, i.e. that it is but a *way* of contemplating objects'. The reader is asked to try to imagine a space.

You will immediately find that you imagine a space for *that* Space to exist in—in other words, that you turn this first space into a *thing* in space: or if

---

[19] Menachem Fisch, 'A Philosopher's Coming of Age: A Study in Erotetic Intellectual History', in *William Whewell: A Composite Portrait*, ed. Menachem Fisch and Simon Schaffer (Oxford: Clarendon Press, 1991), pp. 31–66 (pp. 36–7). See also Immanuel Kant, 'this formal principle of our intuition (space and time) is the condition under which something can be the object of our senses': 'On the Form and Principles of the Sensible and the Intelligible World', in *Theoretical Philosophy 1755–1770*, trans. and ed. David Walford in collaboration with Ralf Meerbote (Cambridge: CUP, 1992), pp. 373–416 (p. 389 (§10)).

[20] William Whewell, *Philosophy of the Inductive Sciences, Founded Upon Their History*, 2 vols (London: Parker, 1840), I, p. 86.

[21] Ibid., I, p. 84.   [22] Ibid., I, p. 84.

you could succeed in abstracting from all thought of Color and Substance, &
then shutting your eyes to try to imagine it—it will be a mere Diagram, and
no longer a construction *in* space but a construction *of* Space.[23]

This is Romantic abstract space: not space imagined as a thing, but
the *condition* for imagining things. Coleridge calls it 'the universal
antecedent Form and ground of all Seeing', and 'seeing' here seems to
mean understanding as well as visualizing.[24] It is 'antecedent' because it
comes before perception.

So we have a concept of space before we perceive things *in* space.
But what kind of space is it that exists as a mental category prior to our
experience of particular spaces in the world? It must be infinite: if it were
bounded it would be only a part of space, not space itself. This leads to a
Kantian insistence on 'universal' or 'absolute' space, which is unitary and
whole. This absolute space contains all other spaces, but is not made up of
them: it exists whether or not they do. It is not the sum of all possible imaginable spaces, because, as Whewell puts it, 'Absolute Space is essentially
one'.[25] Here is another sign of the very great impact of *Naturphilosophie*
on British nineteenth-century science: the Kantian emphasis on the unity
of space contributed to the increasingly dominant ideology of the unity
of natural laws which affected much thinking about the physical sciences
in this period and which seemed to be borne out by the identification
of electricity and magnetism as aspects of the same force in the 1830s.

Whewell is clearly struggling with his terminology for spatial ideas
throughout *The Philosophy of the Inductive Sciences*. He is uneasily aware
of the instability and even inadequacy of his writing about space. These
problems exist because the idea of space is not really expressible in verbal
language: the 'verbal enunciations of the results of the idea cannot
be made to depend on each other by logical consequence; but have a
mutual dependence of a more intimate kind, which words cannot fully
convey'.[26] Language cannot fully express spatial knowledge; and yet a
knowledge of space is part of the language of nature.

---

[23] Coleridge to James Gillman, junior, 22 October 1826, in *Collected Letters of Samuel Taylor Coleridge*, ed. Earl Leslie Griggs, 6 vols (Oxford: Clarendon Press, 1956–71), VI (1971), p. 630.

[24] Ibid., VI, p. 630.

[25] Whewell, *The Philosophy of the Inductive Sciences*, I, p. 85. Compare with Kant's 1768 account of an 'absolute and original space': 'Concerning the Ultimate Ground of the Differentiation of Directions in Space', in *Theoretical Philosophy 1755–1770*, pp. 361–72 (p. 371). It should be noted that Kant later modified this concept of absolute space.

[26] Whewell, *The Philosophy of the Inductive Sciences*, I, p. 91.

The quality that made abstract space so attractive and so necessary an ingredient in the nineteenth-century spatial imagination was the one that made it difficult to write about. Abstract or ideal space is space conceived as pure, beyond human intervention. Was it therefore beyond human language? There was no universal view about the answer to this question in the first half of the century. And yet this kind of space was one of the fundamentals of contemporary science.

Modern popular accounts of science tend to group different kinds of science by disciplines (such as the 'hard', 'life', and 'earth' sciences), or according to whether they emphasize the experimental or the theoretical. Early nineteenth-century equivalents often preferred this latter kind of division. Popularizing texts leant particularly heavily on categorizations of science based on its relationship with the practical. The attempt to draw in a new, wider audience for scientific information meant that the experimental sciences were often presented as an extension of the technological knowledge that would already be familiar to readers employed in manufacturing. But it was still very important to emphasize that technological applications were underlain by a set of scientific laws, and that underlying them in turn were fundamental ideas without which science could not be performed or understood. These ideas supported all scientific thought and all material phenomena, but were themselves abstract rather than material. Among these fundamental ideas was space. But there was no consensus about what space was. Material spaces were easy to describe and define, but abstract space, the idea of space, was not. While many sciences, such as dynamics and electromagnetics, were investigating particular properties of space, or things in space, there remained the problem of what space itself was. The topic might be claimed to be a metaphysical or psychological one. But it also affected work in the physical sciences and had to be dealt with there.

John Herschel took a sanguine, common-sense view of the problem in his 1830 *Preliminary Discourse on the Study of Natural Philosophy*. He argued that the 'existences' of abstract science, including space and time, 'are so definite, and our notions of them so distinct, that we can reason about them with an assurance, that the words and signs used in our reasonings are full and true representatives of the things signified'.[27] But

---

[27] John F. W. Herschel, *A Preliminary Discourse on the Study of Natural Philosophy*, (London: Longman, 1830), pp. 18, 19–20. At the beginning of the *Principia*, Newton made the similar point that time, space, place, and motion required no definition, 'as

others disagreed. Coleridge, for one, insisted that space was one of the most problematic of categories and that those who took the view that there was any easy way out of the difficulty automatically disqualified themselves from understanding *Biographia Literaria*:

[there is] one criterion, by which it may be rationally conjectured before-hand, whether or no a reader would lose his time, and perhaps his temper, in the perusal of this, or any other treatise constructed on similar principles. The criterion is this: if a man receives as fundamental facts, and therefore of course indemonstrable and incapable of further analysis, the general notions of matter, spirit, soul, body, action, passiveness, time, space, cause and effect, consciousness, perception, memory and habit; if he feels his mind completely at rest concerning all these, and is satisfied, if only he can analyse all other notions into some one or other of these supposed elements with plausible subordination and apt arrangement: to such a mind I would as courteously as possible convey the hint, that for him the chapter was not written. [...] For these terms do in truth *include* all the difficulties, which the human mind can propose for solution.[28]

Herschel and Coleridge, in many ways, represent different ends of the spectrum of early nineteenth-century ideas about the relationship of science and language. Herschel emphasizes common intuitive access to meaning through unambiguous language, and Coleridge is disgusted by such an unnuanced and unquestioning attitude. Coleridge's rhetoric demonstrates derisively his contempt for the optimists at the oppos-ite end of this spectrum. While presenting itself as affably generous, Coleridge's list of 'difficulties' ('matter, spirit, soul, body, action, passive-ness, time, space, cause and effect, consciousness, perception, memory and habit') heaps abstract nouns together in a miscellany which, with its rapidity and absurdly easy transitions from one item to the next, mockingly imitates the indiscriminateness of the 'mind completely at rest'. Further mockery is clear in the politeness and scrupulosity of

being well known to all', but he then registered doubt about who was to be included in that 'all': 'the common people conceive those quantities under no other notions but from the relation they bear to sensible objects. And thence arise certain prejudices.' Following this caveat, Newton proceeded to distinguish between absolute and relative space: Isaac Newton, *Mathematical Principles of Natural Philosophy and his System of the World*, trans. Andrew Motte, rev. Florian Cajori, 2 vols (Berkeley: University of California Press, 1966), I, p. 6.

[28] Samuel Taylor Coleridge, *Biographia Literaria; or, Biographical Sketches of My Literary Life and Opinions*, ed. James Engell and W. Jackson Bate, 2 vols (London: Routledge & Kegan Paul, 1983), I, pp. 234–5.

Coleridge's response to the complacent reader, who is treated with patronizing delicacy. Among the list, of course, are many concepts which Herschel would have been very likely to exclude from his list of the objects of abstract science. Coleridge suggests that those who are satisfied regarding any one of the concepts must be so uncritical as to be satisfied regarding all the others. Although they agree that space is one of the 'fundamental facts' of science, they disagree about what that means for science itself and for anyone who wants to study it. The disagreement arises over the relationship of language to scientific investigation: Herschel wants general agreement about meanings, even if the nature of the things named (such as space) are not clearly defined, in order that science can proceed to the next stage. Coleridge does not see how any valid next stage can be proceeded to until the precise meanings of the fundamental terms are clarified. Abstract space is thus a very interesting case study in an important and long-running battle in the relationship of science and verbal culture in the early nineteenth century.

Herschel and Coleridge both believe there to be fundamental concepts of science without which no reasoning can be performed. These concepts necessarily belong to the abstract sciences rather than to the experimental ones: experiment is based on the principles of abstract science. The point is not that these fundamental concepts are inherent in the human mind or in human languages (often quite the contrary), but rather that they are inherent in any abstract understanding of the universe; perhaps that they are inherent in the universe itself. The two writers do not agree, however, about what these fundamental concepts are, and would certainly disagree about how far they were understood.

Herschel's *Preliminary Discourse* describes the objects of abstract science, of which space is one, as including 'those primary existences and relations which we cannot even conceive not to *be*'.[29] He gives the metaphysical status of their 'existence', little attention. The emphatic italics of '*be*' seem to voice his unwillingness to question the nature of that being.

Coleridge, on the other hand, considered that he himself had attended to the problem of the nature of space, even if the result of his attention was only a greater awareness of human ignorance on the subject. In a note of 1807, he compares his own feeling of the mysteriousness of

[29] Herschel, *Preliminary Discourse*, p. 18.

...d other concepts of abstract science, with a Lockean sense that ...oncepts can be clearly and unproblematically understood. He ...guishes implicitly between two kinds of knowledge, one true and ...aps unattainable, one false and promulgated by the ignorant:

> Time, Space, Duration, Action, Active, Passion, Passive, Activeness, Passiveness, Reaction, Causation, Affinity—here assemble all the Mysteries—known, all is known—unknown, say rather, merely known, all is unintelligible / and yet Locke & the stupid adorers of ~~this~~ that *Fetisch* Earth-clod, take all these for granted—[...].[30]

This list includes terms which, from the perspective of the later nineteenth century, belong to two different disciplines: physics (time, space) and chemistry (reaction, affinity). Nonetheless, this early list is more clearly concerned with the non-human sciences than is the equivalent list in *Biographia Literaria* (above, p. 154). Five elements of the 1807 list, including 'space', appear again in the 1817 version. Six of the other elements in the *Biographia* list (spirit, soul, consciousness, perception, memory and habit) are drawn from human-centred disciplines; only 'matter' has reference in abstract science. By comparison, the only item connected with the human in the earlier list is 'passion'. In the decade between the notebook entry and *Biographia Literaria*, then, Coleridge's sense of the 'Mysteries' became much more human-centred, but his underlying argument that these terms and concepts are not 'fundamental facts' but rather fundamental 'difficulties' did not change.

But for many early nineteenth-century writers, 'passion' and the fundamentals of abstract science did not belong in the same category at all. Abstract space, in particular, was attractive, compelling, and had cultural prestige largely because it was beyond the reach of passion, accessible only by reason and intellect. Geometry was the best-known method of reaching this abstract space. Wordsworth famously called geometry, in lines that were unchanged from the 1805 to the 1850 *Prelude*, 'an independent world, | Created out of pure intelligence'.[31] As De Quincey wrote of Wordsworth, 'the secret of [his] admiration

---

[30] *The Notebooks of Samuel Taylor Coleridge*, ed. Kathleen Coburn, 2 double vols (London: Routledge & Kegan Paul, 1957–62), II text (1962), ¶ 3156.

[31] Wordsworth, *The Prelude* (1805) in *The Prelude: 1799, 1805, 1850*, ed. Jonathan Wordsworth, M. H. Abrams and Stephen Gill (New York: Norton, 1979), Book VI, ll.166–7.

for geometry lay in the antagonism between this world of bodiless abstraction and the world of passion'.[32] Geometry offered access to a world beyond the emotional, the autobiographical and the historical. 'Mighty is the charm | Of those abstractions to a mind beset | With images, and haunted by herself': Wordsworth contrasts the besetting 'images' of subjective life with the beguiling 'diagrams' that provide a visible sign of the consolation geometry gives its student.[33] And with this reference to diagrams we pick up one of the controversial topics in early nineteenth-century discussions of geometry. If the truths of geometry are transcendent ones, what role should the senses play in learning and demonstrating them?

As we saw in Chapter 3, geometry and mathematics generally were often argued to be key examples of the kinds of science which did not rely on observation of external phenomena. Yet demonstration of geometry's truths could involve a great deal of visible data in the form of diagrams and other illustrations. Increasing numbers of writers on the subject argued that the teaching of geometry should include as much visual information as possible, and even practical activities. But this argument tended to draw the study of geometry out of the purely rational arena which had been the source of much of its prestige, and into an unpleasingly empirical world in which the transcendent nature of geometry's certainties would be lost. Geometry became a test case for debates about the shifting nature of education in the early nineteenth century, and it was fought over with fervour even by non-specialists, because at stake were crucial decisions about the authority and the purpose of knowledge. As with many carriers of cultural prestige in this period, geometry was profoundly woven into national debates about access to and production of knowledge under new social, political, and technological conditions. But even more than this, geometry stood for a kind of knowledge that went beyond any merely human epistemology: so rational, systematic, and certain that for some writers it had, paradoxically, an almost religious significance. To tamper with that certainty by watering down the rationality of geometry in favour of empirical methods of teaching designed at making it relevant and accessible to the lower classes (or

---

[32] Thomas De Quincey, 'Lake Reminiscences. II.' (1839), in *The Works of Thomas De Quincey* gen. ed. Grevel Lindop, 21 vols (London: Pickering and Chatto, 2000–3), 11 (2003), 66–91 (80).

[33] Wordsworth, *The Prelude* (1805), VI, 158–60; 152; (1850), VI, 151.

others) was—to some commentators—an appalling undermining of the foundations of culture.

GEOMETRY AND ACCESS TO ABSTRACT SPACE

A great deal of early nineteenth-century rhetoric about education and learning proposed geometry as the purest and most rational kind of knowledge and imbued it with moral as well as intellectual qualities. As Joan L. Richards puts it, 'discussions of man's intellectual aspirations and limits consistently referred to [geometry] as a pivotal example'.[34] Geometry was the foundation of the sciences not only because it dealt with space, one of the fundamental elements of abstract science, but because it inculcated the reasoning process which was the only secure means of producing and verifying knowledge. To learn geometry according to the system propounded by Euclid trained the mind in rigorous and abstract reasoning and was thus essential for any further study of scientific or philosophical subjects—or so many professors of mathematics argued. The example of Michael Faraday, who lacked a university education and had surprisingly little mathematical skill (he said rather wistfully 'all my mathematics consist in that rough natural portion of geometry which every body has more or less') shows that—contrary to the argument that geometry and mathematics more generally were a sine qua non for scientific understanding—at least for the first few decades of the century, it was possible to excel in science without a classical geometrical training.[35] Nonetheless, the vital importance of geometry continued to be widely asserted, and not only for the educational and scientific elite: industrialization meant that the application of scientific principles became important to men without access to advanced education, and accordingly self-help books on geometry emphasized the practical benefits as well as the edifying nature of familiarity with the subject.

Dionysius Lardner, Professor of Natural Philosophy and Astronomy in the University of London, contributed a volume on geometry to the *Cabinet Cyclopaedia*, one of the great manifestations of the artisan education movement:

[34] Joan L. Richards, *Mathematical Visions: The Pursuit of Geometry in Victorian England* (San Diego, Calif.: Academic Press, 1988), p. 3.
[35] Faraday to William Whewell, 19 September 1835, in *Correspondence of Michael Faraday*, II, p. 278.

The advantages [...] accruing to artisans from a due cultivation of the principles of pure geometry, are not confined merely to invigorating their discursive powers, nor to storing their memories wirh principles of art useful in almost every department of their daily occupations; but in addition to these important purposes, such a study inspires a taste for that precision of construction, and a love for that accuracy of form, in the absence of which no artisan or engineer, whatever be his grade, can hope to arrive at great professional excellence.[36]

So high was geometry's standing in early nineteenth-century culture that 'Euclid' could be used to indicate any system of knowledge based on rigorous and connected principles. William Whewell's *Mechanical Euclid* (1837) was a textbook on mechanics and hydrostatics and had nothing to do with geometry, but used Euclid's name and the format of his *Elements* (definitions, axioms, and postulates) in an attempt to give the fairly new sciences Whewell was expounding some of the cultural capital of geometry: 'by calling this little work The Mechanical Euclid, I mean to imply, that I have aimed at making it such a coherent system of reasoning, as that of which Euclid's name is become a synonym'.[37] And Euclid's method could be invoked to support arguments that had nothing to do with geometry and were right outside the mathematical sciences altogether: Coleridge compared his approach in *Aids to Reflection* with that of the *Elements*, shoring up the validity of his moral arguments with the cultural prestige of Euclid: 'we have begun, as in geometry, with defining our Terms; and we proceed, like the Geometricians, with stating our POSTULATES; the difference being, that the Postulates of Geometry *no* man *can* deny, those of Moral Science are such as no *good* man *will* deny.'[38]

As Coleridge's application of Euclid's kudos to bolster his 'science' of morality indicates, geometry served as a kind of gold standard for the emerging sciences. It was especially important at this particular moment in the history of scientific culture that such a standard exist and be recognized as such. Loud claims to authority were being voiced by competing disciplinary and organizational systems, all seeking to

[36] Dionysius Lardner, *A Treatise on Geometry, and Its Application in the Arts* (London: Longman *et al.*, 1840), p. 15.

[37] William Whewell, *The Mechanical Euclid, Containing the Elements of Mechanics and Hydrostatics Demonstrated after the Manner of the Elements of Geometry [...]* (Cambridge: Deighton; London: Parker, 1837), p. v.

[38] Samuel Taylor Coleridge, *Aids to Reflection*, (1824), ed. John Beer (London: Routledge and Princeton, N.J., Princeton University Press, 1993), p. 136.

order current knowledge in such a way as to stimulate the production of new knowledge, but failing to agree on the clearest and most useful way of presenting information as a system. From the organization of topics in encyclopaedias to university curricula to self-help textbooks, the structure of knowledge was up for grabs in early nineteenth-century Britain. It was crucial that information be organized in such a way as to show the connections between different fields, both so that learners could understand how one science developed from another, and so that experts might perceive ways in which the sciences could be (re)united. But the days in which a man might claim at least acquaintance with all branches of learning were unquestionably over, and so specialization must not be inhibited by organizational principles that emphasized breadth at the expense of depth. Geometry was frequently presented as the uncontentious model for excellence in the organization of knowledge. And until at least the mid-nineteenth century, 'geometry' for many commentators meant, broadly, Euclid. Dionysus Lardner was voicing a mainstream view when he wrote that 'the books of Euclid and their propositions are as familiar to the minds of all who have been engaged in scientific pursuits, as the letters of the alphabet. The same species of inconvenience, differing only in degree, would arise from disturbing this universal arrangement of geometrical principles as would be produced by changing the names and powers of the letters.'[39]

But assertions of geometry's uncontentious supremacy were partially attempts to silence a debate about the validity of the structuring principles of Euclid's *Elements* in particular and the role of abstract mathematics in general within the new knowledge economy. Although the development of non-Euclidean geometries by Nicholai Lobachevskii and János Bólyai began in the 1820s, it took around a half-century for mainstream British mathematics teaching to mount a successful challenge to the preeminence of Euclid in geometry.[40] During this time, however, numerous attempts were made to replace the *Elements* with more modern and practically-minded systems, and increasingly the centrality of geometry of any kind in a worthwhile education was being questioned. William Whewell, Dionysius Lardner, and the other proponents of classical geometry had a fight on their hands over the importance of geometrical and mathematical studies in the production

[39] Dionysius Lardner, *The First Six Books of The Elements of Euclid [...]* (London: Taylor, 1828), p. xii.
[40] Richards, *Mathematical Visions*, pp. 66–7.

and acquisition of knowledge. Some, like Lardner and Brougham, tried to win this fight by extending knowledge of and belief in geometry to the emerging artisan and manufacturing class. Others, like Whewell and (though he did not take part in the debate in the same way as Whewell) Coleridge, argued that geometry was an equally useful basis for all the sciences and for all intellectual pursuits of any kind because it was in some sense an expression of principles already existing in the human mind. For these writers, learning geometry was, more than any other study, learning reason itself. A writer in the *Prospective Review* summed up this widely expressed view when he remarked in 1847 that 'the Elements of Euclid seem unquestionably to afford the best initiatory exercise of the reasoning faculty. [...] This lucid development of facts is undoubtedly a wholesome exercise to the mind, more especially of a beginner, and there is no reason to suppose that the habits of thought thus acquired are incapable of being applied to other branches of knowledge.'[41] And the fact that this journal was reviewing a new geometry textbook at all is an indication of the centrality of Euclid in early nineteenth-century culture.

GEOMETRY AND THE INDIVIDUAL

Although Coleridge's views on access to and production of knowledge were at variance with those of many of the other writers discussed in this book, nonetheless he was very much of his time in his veneration of geometry. He seems to have conceived space very largely as geometry, and to have believed that geometry was crucial in almost all branches of thought.[42] This was because geometry was for Coleridge (as for Whewell) not an arbitrary collection of knowledge but a systematization of inherent mental qualities. In *The Friend*, for instance, he treats geometry as part of the inner sense: in order to understand geometry one needs neither experience nor any of 'those other qualities of the mind which are so differently dispensed to different persons, both by nature and education'. Instead, 'all men in their senses possess all the

---

[41] *The Prospective Review, A Quarterly Journal of Theology and Literature*, 3 (1847), pp. 108–11, anonymous review of Robert Potts, *Euclid's Elements of Geometry* (London: Parker, 1847) (pp. 109–10).

[42] Trevor H. Levere notes that Coleridge shared this considerable stress on geometry with Schelling: *Poetry Realized in Nature: Samuel Taylor Coleridge and Early Nineteenth-century Science* (Cambridge: CUP, 1981), p. 127.

component images, viz. simple curves and straight lines', but he admits that the ability to concentrate on the manipulation of these component images and their logical relations is unevenly distributed through the population.[43] Though some people may be better equipped to exercise their geometrical imagination than others, nonetheless all are possessed of such an imagination.

This belief that geometry is an inherent kind of knowledge, one that everyone has from childhood, stemmed from one of the great sources of geometry's cultural prestige: Plato's account of it in *Meno*. In this dialogue, Socrates shows that 'we do not learn, and that what we call learning is only a process of recollection' by quizzing an uneducated slave boy about geometrical figures. A process of detailed questioning leads the boy to give satisfactory answers about lengths of lines and areas of squares, despite never having learned any formal geometry. Socrates moves from claiming that this shows that 'if he did not acquire this knowledge in this life, then clearly he must have had and learned it at some other time' to the proposition that 'if the truth of all things always existed in the soul, then the soul is immortal'.[44] The sociologist Sal Restivo's wider observation that pure mathematics, including geometry, functions 'in some cases as a source of assurance about an eternal after-life' would certainly seem to be supported by this argument, which was important enough to Whewell that he reminded a Royal Institution audience of it, though he did not expressly agree with it.[45] For Whewell, Plato's understanding that geometry provided 'a certain and solid truth' was one of the pillars of Western civilization, since the recognition that some truths are utterly reliable and universal allowed Socrates and Plato to generate a critique of the Sophists and to found a philosophical school which aimed at achieving some of the certainty of geometry in other fields of thought.[46] 'Since man can know, certainly and clearly, about straight and curved in the world of space, he ought to know,—he ought not to be content without knowing,—no

[43] Coleridge, 'The Landing-Place, Essay V', in *The Friend*, ed. Barbara Rooke, *Collected Works of Samuel Taylor Coleridge*, 4, 2 vols (London: Routledge & Kegan Paul, 1969), I, pp. 159, 159–60.

[44] Plato, *Meno*, in *The Dialogues of Plato*, trans. Benjamin Jowett, I, pp. 247–92 (pp. 270, 275, 276).

[45] Sal Restivo, *Mathematics in Society and History: Sociological Inquiries* (Dordrecht: Kluwer, 1992), p. 143. William Whewell, 'On the Influence of the History of Science upon Intellectual Education', in *Lectures on Education: Delivered at the Royal Institution of Great Britain* (London: Parker, 1855), pp. 3–36 (p. 12).

[46] Ibid., pp. 9–11.

less clearly and certainly, about right and wrong in the world of human action.'[47]

Whether or not geometry had its psychological or spiritual origins in eternity, it was acknowledged to have had its historical origins in very material events and activities. Many nineteenth-century writers retold the story of how the annual flooding of the Nile delta stimulated ancient Egyptians to devise a means of measuring land because, as Lardner put it, the water 'obliterated the ordinary boundaries by which the land was subdivided and appropriated, covering the surface with mud'. The systematized study of geometry thus began as a way of establishing borders and property rights, a role it was to play metaphorically in early nineteenth-century discussions about intellectual borders and rights. Lardner makes a move typical of most writing about geometry in this period when he goes on to say that from this muddy beginning, it 'rose to more noble objects'.[48] Despite its origins, geometry was noble because of its independence of material facts and information supplied by the senses. In accounting for this independence from materiality, Lardner and Whewell, two of the great contributors to this debate, took different paths. Lardner thought that the individual mind progressed from material objects to more abstract notions based on them. Geometrical terms 'signify certain abstracted notions or conceptions of the mind, derived, without doubt, originally from material objects by the senses, but subsequently corrected, modified, and, as it were, purified by the operations of the understanding'.[49] Whewell, on the other hand, believed (with Kant) that geometry is inherent in the mind in its abstract form, because 'the idea of space' is a basic mental category which does not need to be reinforced by practical material experience.[50] Nonetheless, it was vital that this natural understanding of geometry be expanded and rationalized by means of training, without which, for Whewell, no person could claim to be educated: 'among ourselves, and in every other country of high cultivation, no education is held to be raised on good foundations which does not include geometry,—*elementary* geometry, at least'. Geometry is a necessary component of a liberal education.[51]

---

[47] Ibid., p. 11.   [48] Lardner, *A Treatise on Geometry*, p. 1.

[49] Lardner, *The First Six Books of the Elements of Euclid*, p. xiv.

[50] Whewell, *Mechanical Euclid*, p.155.

[51] Whewell, 'On the Influence of the History of Science', pp. 14–15. See also Whewell's *Thoughts on the Study of Mathematics, as Part of a Liberal Education* (Cambridge: Deighton; London: Whittaker, 1835).

Coleridge would have agreed:

Nothing can be rightly taken in, as a part of a liberal Education, that is not a Mean of acquainting the Learner with the nature and laws of his own Mind—as far as it is the representative of the Human Mind generally—[...] the whole of Euclid's Elements is but a History and graphic Exposition of the powers and processes of the Intuitive Faculty—or a Code of the Laws, Acts and ideal Products of the pure Sense.[52]

I discussed earlier how Coleridge asks his reader to conduct a mental experiment intended to convince him that space 'belongs to the mind itself, i.e. that it is but a *way* of contemplating objects'. He asks James Gillman Jr. to try to imagine a space: 'you will immediately find that you imagine a space for *that* Space to exist in—in other words, that you turn this first space into a *thing* in space'. Coleridge goes on to make an analogy with language:

it is the fundamental Mistake of Grammarians and Writers on the philosophy of Grammar and Language that words and their syntaxis are the immediate representatives of *Things*, or that they correspond to *Things*. Words correspond to Thoughts; and the legitimate Order & Connection of words to the *Laws* of Thinking and to the acts and affections of the Thinker's mind.[53]

For Coleridge, language is a kind of mirror of psychology. Geometry, too, shares grammar's ability to express and reflect the laws and principles of the human mind. This belief that the processes of the mind are congruent with the structures of geometry and hence of the natural world conceived abstractly was part of Coleridge's debt to *Naturphilosophie,* which stressed the likeness of human mental activity to the activities of the universe.

For Coleridge this likeness of the mind to the universe is illustrated by the way that geometry tells abstract truths about the world and seems to be a way of thinking that is built into the brain independent of history or culture. This is not to say that no mental effort is involved in thinking geometrically. Will is necessary, as well as pure sense. In his 'Treatise on Logic', for example, a geometrical understanding of visible phenomena is initially the result of a kind of 'sense', but by the end of the passage this 'sense' has become an act, presumably a willed one:

[52] Coleridge to James Gillman, junior, 22 October 1826, in *Collected Letters of Samuel Taylor Coleridge*, VI, p. 630.
[53] Ibid., VI (1971), p. 630.

When connecting 2 bright stars, the one directly above the other extremities of the same line I seemed to have a something between a sense and a sensation of length more perfect than any actual filling up of the interval by a succession of points in contact would have given me. I seemed to find myself acting as it were in the construction of that length undisturbed by any accompanying perception of breadth or inequality which must needs accompany all pictures of a line ... in other words it was a self-conscious act snatched away as it were from the product of that act.[54]

When Coleridge says that he could construct a length or line 'undisturbed by any accompanying perception of breadth or inequality', he means that he had the feeling that he was able to imagine the ideal Euclidean line, length without breadth, which can never be seen. Where does this mental ability to imagine such a line come from? Do we learn it by experience, from being taught geometry? Coleridge thought not: geometry 'supplies philosophy with the example of a primary intuition, from which every science that lays claim to *evidence* must take its commencement. The mathematician does not begin with a demonstrable proposition, but with an intuition, a practical idea'.[55] This is why the will is involved in the perception of an ideal geometrical line: in that act of the imagination, the mind is reaffirming part of its own existence. Similarly, Coleridge recorded in a notebook entry: 'in all inevitable Truths, e.g. that the two sides of a triangle are greater than the third, I feel my will active: I seem to *will* the Truth, as well as to perceive it'.[56] I give this sentence as it is quoted by Owen Barfield, but as he himself points out, Coleridge, writing these lines in his notebook, did not use the word 'triangle', but instead drew a three-sided figure.[57] In a Coleridgean spirit, we should avoid calling this figure 'a triangle', since as he insists in *Philosophical Lectures*, 'I need not I am sure tell you that a line upon a slate is but a picture of that act of the imagination which the mathematician alone consults'.[58] The line on a slate is to the ideal Euclidean line Coleridge thought he was able to imagine when looking at the stars as the three-sided figure on a page is to the ideal geometrical

[54] Coleridge, 'Treatise on Logic', British Museum MS, 2 vols, II, 43, quoted in Owen Barfield, *What Coleridge Thought* (London: Oxford University Press, 1971), p. 14. Ellipsis Barfield's.

[55] Coleridge, *Biographia Literaria*, I, p. 250.

[56] The sentence appears in this form in Barfield, *What Coleridge Thought*, p. 14. It is quoted from *The Notebooks of Samuel Taylor Coleridge*, I text (1957), ¶ 1717.

[57] Barfield, *What Coleridge Thought*, p. 195, n. 3.

[58] *Philosophical Lectures of Samuel Taylor Coleridge*, ed. Kathleen Coburn (London: Pilot Press, 1949), p. 333; quoted in Barfield, *What Coleridge Thought*, p. 15.

triangle, which exists in ideal mental form. Triangular figures are representations or reminders of that form—not triangles but only pictures of them. Wittgenstein wittily enumerates some of the ambiguities of the triangular figure:

This triangle

can be seen as a triangular hole, as a solid, as a geometrical drawing; as standing on its base, as hanging from its apex; as a mountain, as a wedge, as an arrow or pointer, as an overturned object which is meant to stand on the shorter side of the right angle, as a half parallelogram, and as various other things.[59]

The figure is ambiguous, or at least has different meanings dependent on context. Although the concept of a triangle may be clear in the imagination, the representation of that imaginary figure on the page necessarily involves a loss of that original clarity. For Coleridge, indeed, a geometrical figure on a page is not that figure itself, but only a mnemonic for the definition which constitutes the figure: 'the Circle in the diagram is only a picture or *remembrancer* of the Circle, on which the mathematician is reasoning'.[60]

GEOMETRY AND CLASS POLITICS

Geometry signified for many of its early nineteenth-century proponents the fact that abstract space was independent of sensory experience and even that sensory experience could inhibit a true understanding of it. In this period mathematics held an immensely privileged status in the European concept of education, and at the root of its status lay the classical study of geometry. In the changing educational environment of the first decades of the nineteenth century, geometry was obliged again and again to justify its position at the heart of scientific and other intellectual training and became a subject of profound contestation.

[59] Ludwig Wittgenstein, *Philosophical Investigations*, trans. G. E. M. Anscombe, 2nd edn (Oxford: Basil Blackwell, 1958), p. 200^e.
[60] Coleridge to James Gillman, junior, 24 October 1826, in *Collected Letters of Samuel Taylor Coleridge*, VI, p. 636.

Arguments defending its place at the heart of masculine middle- and upper-class education emphasized both its practical applications and its cultural prestige. Geometry taught pure reasoning, clarified and edified the mind, but was also the basis of useful and necessary techniques in military, manufacturing, and other callings. It was widely described in exalted terms as the purest kind of rational activity because it dealt with the immaterial, though its truths supported all kinds of material things, from the orbits of the planets to the building of reliable bridges. Temple Chevallier, the Professor of Mathematics at Durham University, contributed to the raging debate about the place of mathematics, particularly geometry, in a modern education, with the widely held view that 'over no other part ['of the wide field of Mathematical and Physical Science'] is there spread so pure and luminous an atmosphere of unclouded truth' as over geometry.[61]

But though mainstream, this was far from being an uncontroversial view. Two major strands can be identified in early nineteenth-century writing about geometry (i.e. writing for readers other than specialist mathematicians). The first identified geometry with the fundamental education of the reasoning faculties, regarded its study as all but essential for anyone wishing to engage in scientific pursuits, and combined an appeal to its classical antiquity with a Romantic valorization of its immateriality. The second rejected this position and argued that not only geometry but all higher mathematics were overprivileged in contemporary education and culture. The reasons supporting this revaluation of geometry were often based on questions of class. In between these two positions were more moderate views which broadly supported the study of geometry but sought to divest it of its aura of privilege and inaccessibility by teaching it in such a way as to emphasize practical rather than abstract reasoning (and thus, to the adherents of the Euclidean method, denuding it of most of its benefit to the learner).

One arena in which the debate over the role of geometry and mathematics more generally in mass education was hotly debated was the Mechanics' Institutes. The Mechanics' Institutes movement was founded by—among others—Henry Brougham, and had its heyday in the 1820s and early 1830s. Neither as radical as some of its adherents and many of its enemies would have wished, but yet inciting

---

[61] Temple Chevallier, *The Study of Mathematics as Conducive to the Development of the Intellectual Powers* (Durham: Parker, 1836), p. 11.

alarmist comment from much of the Tory press, Mechanics' Institutes have generally been seen by twentieth-century historians as a flawed compromise between Whig reformism and the ideals of mass education.[62] Nonetheless, early speakers at the Institutes, both in London and the industrial provinces, often took the opportunity to discuss the politics of knowledge in general and access to particular branches of it in particular. Some of these speeches were highly political, others framed in terms of warm encouragement to intellectual effort without explicitly considering the party or class implications of such effort. A clergyman named Edward Craig gave a lecture to the Staines Literary Institution in 1836 in which he expounded the standard middle-class view that 'the way to true and satisfying science is only along the strait line of mathematical precision', and urged his audience (for whom his advice probably came too late) to learn mathematics while they were young, because 'such knowledge is an advantage, for which there is no substitute.'[63] This idea that the knowledge of mathematics, and science more generally, would benefit working men was taken up frequently by speakers on such platforms, and debated often with subtlety and a variety of views. One major crux was whether the benefit to be derived from this study was practical and financial, or moral. W. P. Gaskell's 1835 lecture to the Cheltenham Mechanics' Institution reiterated the point frequently made in other contexts, that there is great pleasure to be had from the study of science, or a science, for its own sake, and criticized the instrumental view of working-class education. 'All for themselves and nothing for mankind, has been the vile spirit in which men have been taught to look upon knowledge as power', Gaskell told his audience, and used the example of geometry to indicate how even abstract and unpractical knowledge could bring delight to a learner, though the tone in which he asserts this suggests that he hardly believed it himself: 'it is a fact that there are men who take a pleasure in investigating the

---

[62] See for instance Harold Silver's classic study *The Concept of Popular Education.* Silver emphasizes the Institutes' 'reluctance to consider social and political issues' and the comparative prosperousness of their membership (pp. 214–15). R. K. Webb's view that the Institute movement was 'a sad failure' is harsh; but Alan Richardson's comment that 'the Mechanics' Institutes, despite a promising beginning, [...] degenerated into vehicles for disseminating middle-class ideology' is hard to deny (Webb, *The British Working Class Reader, 1790–1848: Literacy and Social Tension* (London: Allen & Unwin, 1955, repr. New York: Kelley, 1971), p. 63; Richardson, *Literature, Education, and Romanticism: Reading as Social Practice, 1780–1832* (Cambridge: CUP, 1994), p. 220).

[63] Edward Craig, *A Lecture on the Formation of a Habit of Scientific Enquiry, delivered at the Staines Literary Institution* (Staines: Smith, 1836), p. 18.

properties of a triangle or a circle, without any expectation of rendering their conclusions auxiliary or useful to natural science.'[64] But Gaskell went on to politicize his point, arguing that because scientific knowledge progressed from generation to generation it taught a progressive political stance and a critical attitude to government and other forms of authority. 'A people who have discovered that their ancestors were wrong in their notions of the material world, will ridicule the appeal which Sovereigns may make to their maxims in the moral world. [...] Victory in one realm of thought gives inspiration to attempt conquests in another.'[65]

Alfred Smith, who lectured in 1831 at the Ripon Mechanics' Institute on the interaction of class and science, cited mathematics as one of the barriers that had been put in the path of working men who wished to educate themselves, and went much further than many of his fellow Mechanics' Institutes lecturers in identifying the supporters of these barriers. Where many lecturers confined themselves to a relatively mild dig at the medieval Roman Catholic Church's role in preventing access to knowledge, Smith attacked the modern protectors of British upper-class interests as manifested in Oxford and Cambridge, who supported an outdated and elitist curriculum in order to discourage other classes from attempting to achieve an education: 'An acquaintance with Mathematics and the dead Languages, has been absurdly put in the place of knowledge itself; and, [...] the cause of Science has been much retarded, and injured by the mistake. The Universities have vastly favoured, and profited by the delusion'.[66] Smith steps up a rhetorical gear to make his point about access quite clear, and unsurprisingly reaches for a landscape metaphor:

In this manner, [...] the humbler and more numerous classes of society were separated from the learned by a wide and deep gulph; and almost insurmountable barriers raised between them. In this manner high walls and impenetrable hedges were built and reared round the garden of Science, so that the multitude should never enter in to gather flowers or fruit.[67]

Smith sees the Society for the Diffusion of Useful Knowledge as a great part of the solution to the problem—'they will send hundreds,

---

[64] W. P. Gaskell, *An Address to the Operative Classes, being the Substance of a Lecture Explanatory and in Defence of the Nature and Objects of the Cheltenham Mechanics' Institution [...]* (Cheltenham: Gray, 1835), pp. 3, 5.

[65] Ibid., p. 6.   [66] Smith, *An Introductory Lecture*, p. 14.

[67] Ibid., p. 14.

nay thousands,—of a new set of labourers into the boundless and half-cultivated fields of science, to explore new tracts, find new riches, and add to the heap of existing knowledge'—and Brougham, the SDUK's founder, as 'the first of those master-minds who have given an irrestistable impulse to society, which can end in nothing less than enlightening every class of people'.[68]

Brougham and the SDUK promoted a reformist but definitely non-radical educational agenda, and some recent commentators have argued that their efforts tended to hold back the process of making education fully accessible. Alan Richardson notes that the knowledge offered in the first tranche of pamphlets published by the SDUK—hydraulics, pneumatics, and the polarization of light—was anything but 'useful' to most of their likely readers.[69] Richard Johnson's helpful study quotes a series of radical attacks on Brougham and the SDUK for the class interests at work in their policy on usefulness: *The English Chartist Circular* commented, for instance, 'their determination is to stifle inquiry respecting the great principles which question their right to larger shares of the national produce than those which the physical producers of the wealth themselves enjoy'.[70] In place of the curriculum offered by Broughamite reformers, radicals called for knowledge that would be 'useful' in helping working people understand the system that oppressed them; as *The Pioneer* put it in 1834, 'all useful knowledge consists in the acquirement of ideas concerning our conditions in life'.[71] Brougham's original template for the SDUK's publications, laid out in his 1825 *Practical Observations upon the Education of the People*, had envisaged a range of subjects including political and economic topics. But the SDUK's two major series, the Library of Useful Knowledge and the Library of Entertaining Knowledge, included almost nothing in these areas. The criticism the SDUK received for this omission was heeded; in 1834, Brougham and Charles Knight, the SDUK's publisher, founded the Society for the Diffusion of Political Knowledge, which was intended to promulgate knowledge of political economy, legal matters, and other such material

---

[68] Smith, *An Introductory Lecture*, pp. 19, 18.

[69] Richardson, *Literature, Education, and Romanticism*, p. 223.

[70] *English Chartist Circular* no. 37, p. 145, quoted in Richard Johnson, ' "Really Useful Knowledge": Radical Education and Working-class Culture, 1790–1848', in John Clarke, Chas Critcher and Richard Johnson, eds, *Working-class Culture: Studies in History and Theory* (London: Hutchinson, 1979), pp. 75–102, (p. 78).

[71] Quoted in Johnson, ' "Really Useful Knowledge" ', p. 84.

among the working class. But this Society and its publications, projected and actual, sank within a few years.[72] The SDUK later made attempts to provide the beginnings of an explicitly political element to its curriculum: in 1839 and 1840 the Society published two treatises by Brougham on politics and political philosophy, but it was too late: the Society's reputation for avoiding political information was formed.

This question of the 'usefulness' of knowledge was key to the problem of the place of mathematics and geometry in particular in the new educational economy of the early nineteenth century. At the radical end of the political spectrum, many argued that the 'use' of the newly offered scientific knowledge was to support the class interests of the establishment rather than to educate the workers to whom it was directed. George Wheatley, a Benthamite and evidently a powerful speaker, addressed the Whitehaven Mechanics' Institution in June 1832, the month of the passing of the Reform Bill, and attacked the emphasis on science in Mechanics' Institutes and the SDUK as a barrier to genuinely useful knowledge:

In Mechanics' Institutes, science [is] held out as the only becoming, the only needful, object of study—by whom? By Henry Brougham.

And science, the only object inculcated by the Society [for the Diffusion of Useful Knowledge], as appears by their first treatise, explaining the nature and tendency of their views; and of those views, science was the beginning, the middle, and the end—and this first treatise written by whom?—By Henry Brougham.

Now here is design, [...] worldly and base, insidiously working its own ends, under the wretchedly futile pretence of giving the labouring classes some knowledge, making it all in all, in order the more effectually to withdraw their minds from all contemplation of morals and of politics.[73]

Instead, Wheatley proposed educational provision that would teach men things that pertained to their trades, and then things that would make them good citizens and good men. Particularly, there should be plenty of literature in the libraries of the Mechanics' Institutes, so that readers could develop their moral and political views. But even this most unusually outspoken Institute lecturer acknowledged the importance

---

[72] R. K. Webb's classic study, *The British Working Class Reader, 1790–1848*, pp. 92–3, gives a succinct account of the SDPK's brief life.

[73] George Wheatley, *An Address on the Necessity, Uses, and Advantage of Affording to the Labouring Classes, the Means of Acquiring General, Scientific, Moral, and Political Knowledge [...]* (Whitehaven: Cook, 1832), pp. 6–7.

of studying mathematics, citing John Locke's view that mathematics taught clear reasoning.[74] Smith and Wheatley, then, with very different political views, agreed that the establishment used an irrational emphasis on certain kinds of knowledge to keep people away from other kinds, but they disagreed about which kinds of knowledge were being used in this way.

Anxiety about the role of geometry in particular and science in general was also being expressed in class terms at quite the opposite end of the social spectrum. For example, the Savilian Professor of Geometry at Oxford University, Baden Powell, was deeply concerned about the absence of proper scientific study from Oxford's curriculum. If the educated elite in Britain knew no science they would be overtaken by those who did:

> Scientific knowledge is rapidly spreading *among all classes* EXCEPT THE HIGHER, and the consequence must be, that that Class *will not long remain* THE HIGHER. If its members would continue to retain their superiority, they must preserve a real *preeminence in knowledge*, and must make advances at least in proportion to the Classes who have *hitherto* been below them.[75]

The tenor of Baden Powell's work is misrepresented by this apparently inflammatory quotation; in fact the lecture from which this passage comes is clear-sighted, practical, and humane, dealing only with Oxford students, and recognizing that for most of them, forced engagement with Euclid was no substitute for learning some of the natural sciences: 'if a considerable proportion of the Undergraduates go through a portion of the Elements of Euclid, it is chiefly by compulsion, without an idea beyond the letter of the text, and this often learned by rote'.[76] Instead of reifying geometry and allowing it to stand for all science, Baden Powell argued that it should be seen as only one branch of mathematics, and not one that was important for studying every scientific discipline: 'The Physical Sciences may be pursued to a very considerable extent with scarcely any introduction of Mathematics; and even when Mathematical reasoning is applied, such a knowledge of it as supposes the learner to be perfectly at home in the letter of Geometrical demonstrations, is far from being indispensable to following up many very interesting

---

[74] Wheatley, *An Address*, p. 12.

[75] Baden Powell, *The Present State and Future Prospects of Mathematical and Physical Studies in the University of Oxford* […] (London: Parker, 1832), p. 27.

[76] Ibid., p. 10.

portions of such investigations, and the study of those laws which regulate some of the most remarkable phenomena of the material world.'[77]

Baden Powell's attempt to instate a scientific curriculum in which geometry would not be considered a necessary foundation brought to the elite an argument that had for some time been made about the education of the lower classes. Neil Arnott's *Elements of Physics* was an influential example of this argument. Arnott, a Benthamite, a member of the Royal College of Physicians, and the inventor of the water-bed, had a very liberal view of what he called 'the *diffusion of existing knowledge* among the mass of mankind', which through a free press was leading to free political institutions and 'the extraordinary moment of revolution or transit, in which the world at present exists!'[78] But part of this diffusion of knowledge had been held back by the insistence on the need to learn higher mathematics, which was for most people irrelevant as well as rebarbative: 'The mathematical knowledge, acquired by every individual in the common experience of childhood and early youth [...] is sufficient to enable students to understand all the great laws of nature,—nearly as the knowledge of language obtained at the same time is sufficient, without any study of abstract grammar, to enable him to converse on all common subjects.' Geometry was singled out as an example: 'there are few persons in civilized society so ignorant, as not to know that a square has four equal sides, and four equal corners or angles, or that every point in the circumference of a circle is at the same distance from the centre'.[79] Arnott's idea was that geometrical knowledge was acquired, not inborn (as Plato, Kant, Whewell, and Coleridge in their various ways thought), and his examples show again how different ideas could be during this period about what constituted 'geometry'. To know that a square has four sides is only to know the meaning of the *word* 'square'. Arnott's example of knowledge about a circle, similarly, is just a restatement of Euclid's *definition* of a circle. Neither of these is knowledge that could properly be called geometrical. Rather, they are linguistic. It is not surprising that Arnott goes on to identify language as the major contemporary problem

---

[77] Ibid., p. 16.
[78] Neil Arnott, *Elements of Physics, or Natural Philosophy General and Medical, Explained Independently of Technical Mathematics, and Containing New Disquisitions and Practical Suggestions*, 3rd edn (London: Longman *et al.*, 1828), pp. xvi–xvii.
[79] Ibid., p. xxxii.

slowing down the diffusion of scientific knowledge: 'barbarous names' have 'rendered [mathematics] terrible to the great mass of mankind': '*Arithmetic, algebra, fluxions, geometry, mathematics,* &c., deters common minds from the study: but men of talent are now smoothing the access, by translating the old technicalities into the common languages'.[80] Nonetheless, the optimism of Arnott's argument made it useful to popularizers of science. The opening number of *The Penny Mechanic*, for instance, approvingly cited Arnott's belief that 'the mathematics of common sense' was all that was necessary to investigate the laws of nature.[81]

At this point it is clear how widely separated early nineteenth-century views about the role of geometry in education were. For the patrician Romantics, such as Whewell, a fundamental understanding of geometry was inherent in the human mind, but geometry was an essential part of education because it taught a standard of truth to which all other knowledge should aspire. For the reformists, such as Baden Powell, geometry was *not* inherent in the mind at all; indeed he was prepared to accept the suggestion that 'there are individuals whose minds are so constituted, that they are physically incapable of the degree of abstraction of thought necessary for going through Geometrical demonstrations'.[82] Writers whose circumstances, audiences and politics were as different as those of Arnott and Baden Powell shared a belief in the development of a scientific curriculum in which geometry was not the gatekeeper, but surrendered its claim to a uniquely privileged status in general education. Under these conditions, abstract space, the kind of space dealt with by geometry, seemed to be under threat not only of eradication from the educational experience of increasing numbers of people but also of diminution into physical space, the kind of space which could be investigated with tools.

At the back of William Ritchie's 1833 textbook on geometry is an advertisement: 'N.B. The author being firmly convinced of the vast importance of the simple INSTRUMENTS described in this volume, has made arrangements that teachers or families may be supplied with

[80] Arnott, *Elements of Physics*, p. xxxiii.

[81] *The Penny Mechanic: A Magazine of the Arts and Sciences*, 1, no. 1 (5 November 1836), p. 2.

[82] Powell, *The Present State*, p. 15.

them on application to the Publisher.'[83] Plato's ideal of geometry involved nothing more than diagrams scratched with sticks in sand. The use of protractors and compasses indicated that geometry was losing its abstractness and that its spaces were becoming democratized and concretized and—to the opponents of this development—fundamentally denatured.

[83] William Ritchie, *Principles of Geometry Familiarly Illustrated, and Applied to a Variety of Useful purposes, Designed for the Instruction of Young Persons*, (London: Taylor, 1833), n. p.

6

Bodies in Space:
Ether, Light, and the Beginnings
of the Field

The problem of how bodies act on one another—how forces are transmitted from one body to another—was one of the key questions of early nineteenth-century science. It was at the heart of debates about the nature of heat, light, electricity, gravity, and magnetism; it was important in attempts in physiology to account for the connection between mind and body; and it spilled out into the philosophy of science in its implications for materialist explanations of natural phenomena. More than the search for origins, which in many fields was still considered (publicly at least) to be the province of theologians, the search for the means of connection between objects and forces shaped early nineteenth-century science.

Science's deep interest in means of connection was a manifestation of a wider cultural concern. The questions raised by the spread of literacy, access to print and information through new networks reaching new readers in early nineteenth-century Britain were crucially ones of transmission. Equally, the aspects of contemporary knowledge production and dissemination I have discussed in earlier chapters can also be understood as problems in transmission. Debates about how best to sort and classify knowledge and then how—or whether—to regulate access to it, were premised on beliefs about the value of knowledge moving from one kind of space (disciplines, books, minds) to another, and about what happened to knowledge when such movements took place. And while these questions were being debated in political, literary, and journalistic writing, contemporary science faced the double problem of building structures to facilitate transmission within and to some extent between its disciplines and negotiating new relationships with the means of transmission of language, ideas, and resources within broader cultural formations. Of course there were strong connections between scientific and literary conceptions of transmission. Clare Pettitt's suggestive work on late Victorian anxieties about originality and transmission, for example, shows that beliefs about the way in which electricity is transmitted posed a challenge to some culturally crucial ideas including the fundamental categories of individuality and property.[1]

This chapter investigates the way in which this widespread concern with transmission was focused in contemporary scientific writing about three ambiguous and problematic entities: ethers, light, and fields. Accounts of each of these entities presented different problems in explaining the transmission of forces across space, and in turn produced different explanations of what space itself is. In each case, the contending theories were necessarily founded on competing descriptions of space, because space was the location in which bodies existed and through which forces were propagated. Changing accounts of space in the physical sciences in the nineteenth century registered profound and fundamental shifts in the understanding of matter and energy.

Traditional accounts of the period, which see it as characterized by increasing materialism, have not acknowledged the profoundly dematerializing tendency in nineteenth-century physical science, which produced a revolution in scientific conceptions of space and matter by the end of the first half of the century. In terms of lived space, most recent accounts of Victorian domestic geography tend, rightly, to emphasize increasing fullness: rooms, houses, and cities were all, in their different ways, filling up with solid objects during the first six or so decades of the century. Judith Flanders is typical of modern historians of Victorian domestic spaces when she argues that from roughly mid-century, 'more and more objects began to accumulate [in the house], becoming a deluge of 'things' [...] There were things to cover things, things to hold other things, things that were representations of yet more things.'[2] Images of spaces full of desirable objects linked homes with warehouses as markers of prosperity in a capitalist economy. Dickens's description of the vulgar fullness of Ralph Nickleby's house in *Nicholas Nickleby* shows the plenitude of material things threatening to breach the boundaries separating the home from the public thoroughfare:

---

[1] Clare Pettitt, *Patent Inventions: Intellectual Property and the Victorian Novel* (Oxford: Oxford University Press, 2004), p. 275.
[2] Judith Flanders, *The Victorian House: Domestic Life from Childbirth to Deathbed* (London: HarperCollins, 2003), p. 151.

The softest and most elegant carpets, the most exquisite pictures, the cost-
liest mirrors; articles of richest ornament, quite dazzling from their beauty
and perplexing from the prodigality with which they were scattered around,
encountered [Kate] on every side. The very staircase nearly down to the hall
door, was crammed with beautiful and luxurious things, as though the house
were brim-full of riches, which, with a very trifling addition, would fairly run
over into the street.[3]

Dickens begins by identifying the objects in Nickleby's house—carpets,
pictures, mirrors—but then abandons these distinctions to call them
simply 'things'. The specificity of this highly processed matter breaks
down. Nickleby intends the excessive fullness of the house to suggest
permanence and stability, but as scenes of household sales in Victorian
novels illustrate, such collections of material objects were easily dissip-
ated. Dickens's reader is expected to be sceptical about the permanence
of Nickleby's 'things'. Given enough time, they will take their place
among the 'coal-dust, vegetable-dust, bone-dust, crockery dust, rough
dust and shifted dust' in the heaps belonging to the successors of the
Golden Dustman in *Our Mutual Friend*.[4] Their coherence as part of a
collection is under threat from the mechanisms of exchange and their
permanence as objects is jeopardized by their comparative fragility. And
during the mid-nineteenth century, even their material solidity was
coming into question.

   This chapter argues that even as increased access to manufactured
goods caused the spaces of nineteenth-century urban life to fill up,
changes in physical science's understanding of the relationship of matter
and forces generated a new conception of space filled with forces
interacting in complex patterns but emptied of solid material bodies.
The first part of the chapter considers early nineteenth-century theories

---

   [3] Charles Dickens, *The Life and Adventures of Nicholas Nickleby* (1839), ed. Mark
Ford (London: Penguin, 1999), p. 229 (ch. 19).
   [4] Charles Dickens, *Our Mutual Friend* (1865), ed. Adrian Poole (London: Penguin,
1997), p. 24 (ch. 2). Kate Flint argues that 'the importance of dust to Victorian
culture lies precisely in [a] capacity to suggest the vastness of imaginative conjecture that
may lie behind and beyond the most apparently mundane: the invisible in the visible'
('Dust and Victorian Vision', in Juliet John and Alice Jenkins, eds, *Rethinking Victorian
Culture* (Basingstoke: Macmillan, 2000), pp. 46–62 (p. 60)); for other explorations
of dust in Victorian culture see Joel J. Brattin, 'Constancy, Change, and the Dust
Mounds of *Our Mutual Friend*', *Dickens Quarterly*, 19 (2002), 23–30 and Ellen Handy,
'Dust Piles and Damp Pavements: Excrement, Repression, and the Victorian City in
Photography and Literature', in Carol T. Christ and John O. Jordan, eds, *Victorian
Literature and the Victorian Visual Imagination* (Berkeley: University of California Press,
1995), pp. 111–33.

of ether, a mysterious and much debated substance which pervaded all space and acted as a medium for transmitting natural forces. The second part focuses on the rise of the wave theory of light, a particular and key instance of the use of ether theories to rethink the nature of matter and force. And at the end of the chapter I have a few speculations about the cultural impact of the beginnings of field theory in Michael Faraday's mid-century physics.

ETHER THEORIES AND THE CONFUSIONS OF SHARED TERMINOLOGY

In order to follow the development of the new ideas about the solidity of matter and the nature of space, it is necessary to recognize the large role played in early nineteenth-century physics by ether, a hypothetical substance which filled all space in the universe but which was undetectable by any instrument.[5] It seemed clear to many late eighteenth- and early nineteenth-century scientific practitioners that forces could move from one body to another only by progressing through some medium. Just as sound is carried by air, and cannot move through a vacuum, other forces were also thought to require a medium to carry them, and for this medium late eighteenth-century science generally relied on a number of 'ether theories'. These hypothesized the existence of special media, 'ethers', which were unlike any other kind of matter in that they were invisible, intangible, and weightless—undetectable except via the forces transmitted through them. Although such hypotheses allowed natural philosophers to solve the problem of 'action at a distance', they created the fresh problem of defining the nature of ether—or ethers.[6]

Practitioners in the physical sciences were not by any means agreed on what ether or ethers consisted of, or indeed whether they existed as a material substance at all. As Geoffrey Cantor and M. J. S. Hodge put it, 'some [conceptions of ether] have been interpreted literally, as truly existing *in rerum natura*; others agnostically, as possible representations

---

[5] Mildly confusingly, ether, the gas which had been used as an anaesthetic since the mid-eighteenth century and which Dr Seward self-administers to salve his broken heart in *Dracula*, was named after the hypothetical fluid in which we are interested here.

[6] Mary Hesse gives an excellent account of action at a distance theories in *Forces and Fields: The Concept of Action at a Distance in the History of Physics* (London: Nelson, 1961).

of real physical processes; yet others strictly as fictions useful in the correlating of sensible phenomena'.[7] This mixture of approaches to the ontology of ether added greatly to its fertility as an object of nineteenth-century cultural investigation; but the difficulty of distinguishing among the slew of theories may have contributed to the surprisingly low profile of ether in modern studies of nineteenth-century culture.

Very broadly speaking, two important tendencies in ether theorizing affected early nineteenth-century science. First, as the quest for connections and even identity between natural forces was increasingly dominating the physical sciences, the idea of a singular general ether came gradually to replace the belief in multiple ethers each supporting a different force. This process is at least partly reflected in contemporary popular and literary writing about ether. The other, much more revolutionary development, which does not seem to have made its way far beyond the boundaries of the physical sciences in this period, was the beginnings of a rejection of ether altogether. Roughly a generation after the end of this book's scope, in 1887, the Michelson-Morley experiments significantly weakened support for the existence of ether, and although ether theories have persisted in greatly constrained form into our own times, conventional history of physics tends to see ether as all but defunct after 1905, a victim of special relativity.[8] It is tempting to see these two developments—unification and eventual rejection—as stages in a kind of clearing up process, so that the history of physics gradually shifts from a model featuring many ethers each serving a different purpose, to one in which a single ether which serves all necessary purposes, and then to an ether-free model in which forces are transmitted without need for a separate transmitting medium.

These changes in ether theories were symptomatic of a shift in physical conceptions of space and matter so momentous that it has been called 'a second revolution in science, hardly less fundamental than the famous one of the seventeenth century, [a revolution which] separates

---

[7] G. N. Cantor and M. J. S. Hodge, 'Introduction: Major Themes in the Development of Ether Theories from the Ancients to 1900', in Cantor and Hodge, eds, *Conceptions of Ether: Studies in the History of Ether Theories 1740–1900* (Cambridge: CUP, 1981), pp. 1–60 (p. 2).

[8] For an account of the Michelson-Morley experiment, see Loyd Sylvan Swenson, Jr, *The Ethereal Aether: A History of the Michelson-Morley-Miller Aether-Drift Experiments, 1880–1930* (Austin: University of Texas Press, 1972). Cantor and Hodge recognize the 'widely held view that ether theories suffered a dramatic and sudden demise with the rise of the special theory of relativity', but suggest that further scholarship would illuminate the persistence of ether theories beyond 1905 (*Conceptions of Ether*, p. 53).

all the scientific ideas and institutions of 1840 from those of 1800'.[9] Another historian of energy physics describes 'the abandonment of the imponderable fluids' as 'one of the most significant developments in the transformation of physics in the early nineteenth century'.[10] This tumult was clearly felt in the broader culture of the period, because this second scientific revolution took place, to a considerable degree, *within* that broader culture.

Because ether and its operations had for decades been freely used as metaphors in British literary writing, changes in scientific ideas about ether came by extension to initiate changes in literature's conceptualization of transmission of other kinds. We can use ether as a kind of gauge of physics' relations with literature and other cultural manifestations during this period, though it is not a straightforward matter to take readings from such a gauge. 'Ether' operated in a wide variety of registers in the early nineteenth century, including poetic diction and specialist—and competing—scientific disciplinary usages. Further, in both literary and scientific contexts the substances it denoted could exist at almost any point on a spectrum of metaphoricity. For these reasons, it can be very difficult to determine how far particular nineteenth-century literary allusions to ether reflect scientific ideas rather than metaphorical diction. This is not simply a problem of modern interpretations: 'ether' seemed—mistakenly—to some contemporary writers to be a concept truly common to both literary and scientific culture. This situation resulted in what Gillian Beer felicitously calls 'delusive accords between different communities of meaning' but also in a reinforcement of the belief that the deepest purpose of both kinds of writing was still a shared one.[11]

From the first English translation of Newton's *Principia Mathematica* and into the nineteenth century, the dominant terms used to indicate the baffling physical nature of ether were 'subtle' and 'elastic'. In the closing sentences of the General Scholium of *Principia*, Newton describes what has generally been taken to be a universal ether:

And now we might add something concerning a certain most *subtle* Spirit [*spiritu quodam subtilissimo*], which pervades and lies hid in all gross bodies; by

---

[9] Cantor and Hodge, *Conceptions of Ether*, pp. ix–x.
[10] P. M. Harman, *Energy, Force, and Matter: The Conceptual Development of Nineteenth-Century Physics* (Cambridge: CUP, 1982), p. 19.
[11] Beer, 'Translation or Transformation? The Relations of Literature and Science', in *Open Fields: Science in Cultural Encounter* (Oxford: Clardon Press, 1996), pp. 173–95, (p. 174).

the force and action of which Spirit [...] light is emitted, reflected, refracted, inflected, and heats bodies [...]. But these are things that cannot be explain'd in few words, nor are we furnish'd with that sufficiency of experiments which is required to an accurate determination and demonstration of the laws by which this electric and *elastic* spirit operates.[12]

This text is from Andrew Motte's 1729 translation of the *Principia*, which was during the mid-nineteenth-century almost the only available English version.[13] The 'electric and elastic spirit' mentioned in the last line seems to have been Motte's addition: Newton's Latin gives no such phrase. But Motte's 'elastic' nonetheless became part of the standard vocabulary of ether discourse. And though Newton's term 'spirit' had by the early nineteenth century dropped out of usage in favour of 'fluid', 'subtle' remained, acting to some extent as a marker of the concessions to metaphysics which ether forced on the experimental scientific method. When Herschel, for example, used 'subtle' in his 1830 *Preliminary Discourse* it was in this cautious way: many of the strange properties of electricity, he argued, were explicable by reference to electricity's consisting 'in a rare, subtle, and highly elastic fluid'.[14] Although both words were in frequent use in ether discourse in the early nineteenth century, however, neither 'subtle' nor 'elastic' had a stable, single, generally agreed meaning.

This instability did not deter literary writers from adopting 'subtle' and 'elastic' for writing about ether. Edward Bickersteth, a prolific writer of hymns who was later to become Bishop of Exeter, described the ether in exactly these terms in his very popular mid-century religious epic *Yesterday, To-Day and For Ever*.[15] At the Creation, God gave a command,

> and lo, a sea
> Of subtle and elastic ether flow'd.
> Immense, imponderable, luminous,

[12] The original Latin General Scholium was published as an appendix to the second edition of the *Principia* in 1713. Isaac Newton, *The Mathematical Principles of Natural Philosophy [...]*, trans. Andrew Motte, 2 vols (London: no pub., 1729). My italics.

[13] Motte's translation was reprinted in 1803, 1846, and 1848; that of his chief rival, Robert Thorp, Senior Wrangler at Cambridge in 1758 and archdeacon of Northumberland from 1792, appeared in 1777 and in a second edition in 1802, but Thorp's translation only covered the first book of the *Principia*.

[14] J. F. W. Herschel, *A Preliminary Discourse on the Study of Natural Philosophy* (London: Longman, 1830), p. 330.

[15] The popularity of *Yesterday, To-Day and For Ever* with Victorian readers is attested by its having been reprinted more than a dozen times before 1902.

> Which, while revealing other things, remains
> Itself invisible, impalpable,
> Pervading space.[16]

Bickersteth's use of the phrase 'subtle and elastic' aligned his theological description of ether with contemporary scientific accounts of it. But by describing the ether as a 'sea', Bickersteth was extending the 'fluid' aspect of ether further into literalism than mainstream contemporary science warranted. In doing so, however, he was exactly within the mainstream of contemporary *literary* writing about ether.

Both scientific and literary writers tended to discuss ether as though it were a liquid. This was liable to confirm the impression that 'ether' meant the same thing to writers across the disciplinary divide, despite the fact that in rigorous scientific discourse it was clear that the liquid-like properties of ether were to be considered as strictly limited and specialized. The differences between liquids and ether in this rigorous sense were disguised for many readers because general or popularizing scientific writing often described ether using such formulations as 'subtle fluid'. The close association in nineteenth-century poetry between ether and water might suggest that some poets had registered, to indeterminable degrees, the 'fluid' model of ether which derived from, but was not really endorsed by, contemporary physical science.

Imagining the ether as a kind of infinite liquid, literary writers, especially poets, frequently took advantage of the opportunity to metaphorize ether into seas. Joseph Cottle, who as well as being a protector of Coleridge was a friend of Humphry Davy, assumes readerly familiarity with the 'ether as sea' metaphor when he describes a cosmic region 'where the stars through ether sail'.[17] Similarly, in Thomas Moore's lyric 'The World Was Hushed', the moon 'sail'd through ether slowly'.[18] 'Ether', as in these examples, often contributed a mysterious or sublime solemnity to non-scientific writing, adding to the word's availability for use in a variety of early nineteenth-century registers, but detracting from its precision in any one of those usages.

The difficulty of tracing the lines of connection between scientific and literary ether-discourses during this period is increased by the fact that

---

[16] Edward Henry Bickersteth, *Yesterday, To-Day, and For Ever: A Poem, in Twelve Books* (London: Rivingtons, 1866), p. 112.

[17] Joseph Cottle, 'The Warning [...] ', in *Malvern Hills, with Minor Poems*, 4th edn (London: Cadell, 1829), p. 192.

[18] Thomas Moore, 'The World was Hushed', in *The Poetical Works*, 10 vols (London: Longman *et al.*, 1840–1), V, p. 242.

the two kinds of writing frequently borrowed one another's authority. Edward Bickersteth used 'subtle and elastic' in *Yesterday, Today and For Ever* to mark the coming together of scientific and religious explanations of cosmic unity; and equally scientific writing—particularly in popular contexts—could draw heavily on the mythologizing, narrative, or poetic possibilities of literary ether-discourse. William Whewell, for instance, in his widely-read 1834 Bridgewater Treatise on astronomy, used the one term, 'fluid', for both gaseous and liquid states of matter, supporting the familiar literary image of ether as a kind of intangible water. Further, he wrote about ether via a sort of creation myth, perhaps using ethereal 'fluid' to recall the 'waters' that cover the earth in Genesis 1.1:

> The solid and fluid matter of the earth is the most obvious to our senses; over this, and in its cavities, is poured an invisible fluid, the air, by which warmth and life are diffused and fostered, and by which men communicate with men: over and through this again, and reaching, so far as we know, to the utmost bounds of the universe, is spread another most subtle and attenuated fluid, which, by the play of another set of agents, aids the energies of nature, and which, filling all parts of space, is a means of communication with other planets and other systems.[19]

Whewell's passives ('is poured', 'is spread') leave room for the orthodox reader to assume God's hand at work in the design of the universe, but in introducing an element of narrative into the description its workings they mythologize ether and its origins. John Heraud's poetic apostrophe to Alcyone, the largest star of the Pleiades—

> Among the pleiads reigning without fear,
> In finest ether, pouring from thy urn
> Light to all planets[20]

—uses a different form and classical rather than Christian allusions but is scarcely less mythologizing in its writing about a liquid ether performing a beneficent role in the universe than is this passage of Whewell's scientific exposition. For some scientific writers, to mythologize an already partly or potentially metaphorical entity might have been an attenuation too far, but for Whewell this universal ether was substantial enough to be able to withstand a degree of mythologization.

---

[19] W[illiam] Whewell, *Astronomy and General Physics, Considered with Reference to Natural Theology*, 4th edn (London: Pickering, 1834), p. 140.
[20] John Abraham Heraud, 'Alcyone', in *The In-Gathering [...]* (London: Simpkin, Marshall, 1870), p. 44.

In his Bridgewater Treatise on astronomy and elsewhere, Whewell seems to have thought of ether more literally than did many of his contemporaries, even to the extent of believing that it affected the Earth's motion round the sun.

This possible effect of ether on planetary motion was one of the cruxes in nineteenth-century ether theories. Describing ether as a sea or a fluid, even a 'subtle' one, meant bringing into play other knowledge about fluids which was not always agreed in the scientific establishment, and which was sometimes not at all appropriate to scientific conceptions of ether. If ether was altogether 'subtle', intangible, imponderable, and without substance, then it would not affect the motion of bodies contained in it. On the other hand, if ether was really analogous to normal fluids, it would indeed have such an effect. Because ether pervaded all space, all the things in the universe, including the planets, must be constantly moving through it. And if ether behaved in some sense like a liquid, then it was to be expected that the ether would resist movement through it, if only very slightly, in the same way as water resists the movement of a ship or a swimmer. The ether would exert drag forces on the bodies moving in it, causing tiny but detectable deceleration in the planets' orbits.

Whewell believed that astronomical observation showed that this was happening. Like a number of his contemporaries, he took the apparent acceleration of Encke's comet, whose period of return of 3.3 years is gradually decreasing, as evidence of ether's tendency to retard the planets.[21] In his Bridgewater Treatise he stated firmly that 'there is a resisting medium, and, therefore, the movements of the solar system cannot go on for ever': eventually ether will retard the earth's movement so much that it will be pulled into the sun and destroyed.[22] Envisaging ether as a partially material substance, he invested it with major theological implications, giving it a role in the history of the beginning of the world (in the passage about God pouring ether over the newly created planet) and in the history of the world's end.[23]

The Bridgewater Treatises, of course, were explicitly a natural theological project, in which such a mixing of scientific and Christian

---

[21] Geoffrey Cantor, 'The Theological Significance of Ethers', in Cantor and Hodge, eds, *Conceptions of Ether*, pp. 135–55 (p. 144).

[22] Whewell, *Astronomy and General Physics*, p. 200.

[23] For brief discussion of Whewell's theological project in this part of his Bridgewater Treatise, see Cantor, 'The Theological Significance of Ethers', p. 145.

doctrines was—in the view of most commentators—appropriate.[24] But Whewell's argument from the existence of a substantial ether to the divine creation of the universe links him with a numerous group of writers who hoped to use or modify scientific ideas about ether to support their theological projects. This was a group with which Whewell would hardly have wished to be associated, and with whose views his own knowledge of advanced scientific methods and results greatly contrasted. But when scientific writers such as Whewell contributed to the blurring of disciplinary boundaries in ether-discourse by adopting, when it suited their rhetorical purposes, elements of the sublime, mythologizing, and metaphorical register of literary writing about ether, they risked giving the impression that mythological or metaphorical narratives about ether were being in some way legitimized by science. It is not surprising, then, that some literary and theological writers seem to have felt that they had a place in the debate about the nature of ether, and by extension, in the debates about the forces that ether supported, particularly light.

THE LANGUAGE OF LIGHT

Mary Somerville wrote in 1834 that 'it is impossible [...] to trace the path of a sunbeam through our atmosphere without feeling a desire to know its nature, by what power it traverses the immensity of space, and the various modifications it undergoes at the surfaces and the interior of terrestrial substances'.[25] Many non-scientists seem to have agreed with her, and not only those poets obeying the fashion for sentimental or album verse on the theme of sunbeams, like Felicia Hemans's 1839 poem which exclaims:

> A joy thou art, and a wealth to all!
> A bearer of hope unto land and sea—
> Sunbeam! what gift hath the world like thee?[26]

---

[24] On the history of the Bridgewater Treatises, see W. H. Brock, 'The Selection of the Authors of the Bridgewater Treatises', *Notes and Records of the Royal Society of London*, 21:2 (1966), 162–79 and Jonathan R. Topham, 'Beyond the "Common Context": The Production and Reading of the Bridgewater Treatises', *Isis*, 89:2 (1998), 233–62.

[25] Mary Somerville, *On the Connexion of the Physical Sciences* (London: John Murray, 1834), p. 172.

[26] Felicia Hemans, 'The Sunbeam', in *The Works of Mrs. Hemans*, 7 vols (Edinburgh: Blackwood, 1839), V, pp. 289–90 (p. 289).

The nature of light became a topic of immense importance in the early nineteenth-century physical sciences and leaked through the still very permeable boundaries of scientific disciplinarity to become an object of contemporary cultural interest. Because it crossed the boundaries of physics, optics, mathematics, and chemistry, as well as extending out into theological debate, it offered scope, as we shall see later, for attempted interventions by non-scientists. Interest in the nature of light was widespread in early nineteenth-century culture; and the shift from a corpuscular to a wave theory of light had major significance for nineteenth-century conceptions of space in both scientific and other cultural milieux.

The most generally accepted explanation of light in Britain towards the end of the eighteenth century was Newton's 'corpuscular' theory, which was founded on the idea that light was made up of minute bodies that were fired out by the sun and travelled through space to the eye. These are 'Newton's particles of light' which Blake compares to grains of sand in 'Mock on Mock on Voltaire Rousseau' (1796).

But in the early nineteenth century an alternative theory was gaining ground against this view of light. The 'undulatory' theory saw light as consisting of vibrations or waves rather than as particles rushing through space. This alternative explanation relied on the existence of ether. If light was transmitted by waves, those waves must presumably be moving in some substance. But since light clearly does move through vacuums, its medium must be something other than air, or any other familiar substance. The medium must be immaterial: an ether.

Analogies were frequently made with sound, which travels via waves in the air. For instance, a 'Mr Goddard, of the Polytechnic Institution' gave a lecture at the Society of Arts in 1838 in which he told his audience that the universe is filled with an ethereal medium which 'may be said to be governed by the same laws as the waves of the sea, or the vibrations of sound'.[27] In fact, the two kinds of waves are very different, because light travels via a transverse wave and sound via what has become known as a longitudinal wave. This point was well established by the time of Goddard's lecture, having been made by a number of contemporary scientific practitioners; nonetheless, the comparison between light-waves and sound-waves continued to be frequently made in textbooks and popularizations.[28]

---

[27] *The Times*, 12 Dec. 1838; Issue 16910; p. 6; col B.

[28] For a contemporary explanation of the difference between the waves, see Herschel, *Preliminary Discourse*, p. 261.

The wave theory of light, with its implied ether, was necessarily in contention with the corpuscular theory. The struggle between advocates of the two theories was intense but comparatively short-lived: as Jed Z. Buchwald puts it, 'by the end of the 1830s only a few diehards remained committed to the emission theory, which had almost completely vanished from the journals of science'.[29] The triumph of the undulatory theory was a kind of *cause célèbre* in contemporary science and one which caught the non-scientific reading public's imagination.

Tracing the ways in which literary writers responded to changes in light theories is very difficult, for many of the same reasons that complicate attempts to distinguish changes in literary writing about ether. The problem is partly that of the confusion of terms. Even for those contemporary theorists and experimenters directly engaged in the debate, confusion over terms was a real danger. The debates between the proponents of the undulatory and the corpuscular theories of light were conducted in a mixture of mathematics and verbal language. Many of the mathematical tools used by the undulationists were imported from Continental mathematics, against which some sections of the English mathematical and physical scientific establishment were prejudiced. Most of these mathematical techniques were highly up-to-date and thus specialized. But while the mathematics were controversial, verbal language was fraught with possibilities for accidental misunderstandings because its vocabulary was, to a large extent, shared by advocates of the two different theories. For instance, Jed Z. Buchwald traces some of the most public and heated controversies in light science to the fact that it was 'extremely difficult for scientists who did not think about polarization in the same way to communicate their experimental results to one another without leaving an inevitable residuum of ambiguity'.[30] And for the interested non-specialist, this problem of ambiguity was compounded by the fact that many of the words involved in discussing the transmission of light were also used in non-technical lexis.

Like 'ether', 'rays', 'beams', 'pencils', and other key terms in the sciences of light also belonged in general and literary vocabularies. All these words continued to have their original meanings through the period of the controversy over the nature of light, often with the effect of keeping the remnants or shadows of outdated scientific ideas

---

[29] Jed Z. Buchwald, *The Rise of the Wave Theory of Light: Optical Theory and Experiment in the Early Nineteenth Century* (Chicago: University of Chicago Press, 1989), pp. xiii–xiv.

[30] Ibid. p. xviii.

current in non-specialist parlance. This resulted in a kind of hybrid language which was full of the disciplinary equivalent of 'false friends': terms which to a non-specialist appeared to have a recognizable referent, but which in fact in specialist discourse denoted quite different technical meanings. *Punch* satirized this kind of hybridity in a punning article of 1843 titled 'Punch's Theory of Light':

Light, as we learn from its name, is without weight, and therefore travels with a velocity which renders it uncommonly warm when it reaches us. [...]

Rays are said to form pencils. These must not be confounded with lead pencils; the former produce light, the latter shade. Reflection is best explained by reference to a glass; but it must be observed that two or more glasses, although they often double the effect of light, yet are found to diminish reflection.[31]

The pomposity of some of the phrasing here—for instance, 'it must be observed'—sends up the genre of scientific popularization, but to make the detailed jokes about rays, pencils, and glasses worthwhile, the writer evidently assumed that a fair proportion of *Punch*'s readership would be aware of at least some of the technical terms associated with optics. The fact that the piece is titled a 'Theory' of Light further suggests that many readers would know that the nature of light was the subject of debate, though by 1843 that debate was almost entirely in the past. As the *Punch* article suggests, though, the debate itself continued to be a tool in science's self-understanding and self-presentation, and thus a matter of ongoing public discussion.

By mid-nineteenth century, the adherents of the corpuscular theory of light had decisively lost the battle. Among its high-profile supporters, only the Scottish physicist Sir David Brewster and the English politician and writer on science Henry Brougham continued to resist the undulationists.[32] As late as 1850, at the age of seventy-two, Brougham attacked the wave theory of light in a lecture at the Académie des Sciences in Paris before an audience including François Arago, himself a very distinguished undulationist. But by 1850 Brougham's cause was so hopelessly lost that Arago needed only to make the mildest reply—that

---

[31] *Punch, or the London Charivari*, 5 (1843), 44. 'Pencils' in this context are groups of rays of light converging on or diverging from a single point.

[32] Brougham's opposition to the wave theory of light dated from the very first years of the century; he famously wrote a misguided and very hostile article on Young's work, in which he called his *Bakerian Lecture on the Theory of Light and Colours* 'destitute of every species of merit': *Edinburgh Review*, 1:2 (Jan 1803), 450–6 (p. 450). Michael Faraday objected to the wave theory of light because he rejected the hypothesis of the ether, but he did not publicly support the corpuscular theory.

he believed he could explain Brougham's findings 'by another theory which some philosophers supported'.[33] All the heat had been taken out of the battle over the nature of light.

The clash between the corpuscular and undulatory theories of light had been a clash between mechanical and dynamic understandings of the universe. It was also a clash between two distinct methods of doing mathematical physics. Many modern historians argue that a decisive element in the eventual victory of the undulatory theory was that unlike the corpuscular or 'emission' theory, it had received the attention of some of the brightest mathematicians, using the most up-to-date analytical mathematics; it had in fact become a showcase for the power of the new mathematical methods.[34] This point was acknowledged by contemporary commentators. Baden Powell, the Savilian Professor of Geometry at Oxford and a powerful voice on the undulatory side of the argument, agreed with Herschel that 'if as high mathematical ability had been devoted to the development of the corpuscular theory, as has been to the undulatory, the former might probably have been exhibited in a form equally complete and satisfactory'.[35] This had not happened, though, and instead the powerful and exclusive new tool of Continental analytical mathematics had been devoted to the undulatory theory.

Despite this deployment of advanced foreign mathematics, the topic of the nature of light attracted a great deal of contemporary attention from non-specialists, many of whom were outside the world of mainstream science altogether. One reason for this wide interest was that light—unlike, say, sound or magnetism—was so profoundly associated with theological accounts of the universe. Another factor was the accessibility of the basic equipment needed for optical experiments. Thomas Young's seminal investigations of the nature of light had

[33] *The Times*, 24 Jan. 1850; Issue 20394; p. 6; col A.

[34] For discussion see G. N. Cantor, *Optics after Newton: Theories of Light in Britain and Ireland, 1704–1840* (Manchester: Manchester University Press, 1983), pp. 148–50; see also Jack Morrell and Arnold Thackray, *Gentlemen of Science: Early Years of the British Association for the Advancement of Science* (Oxford: Clarendon Press, 1981) pp. 138–54 and Harman, *Energy, Force and Matter*, p. 25: 'The wave theory of light […] exemplified [Herschel's, Whewell's and George Biddell Airy's] belief in the application of mathematics to physical problems. […] the supporters of the wave theory stressed its mathematical sophistication'.

[35] Baden Powell, *A General and Elementary View of the Undulatory Theory […]* (London: Parker, 1841), p. xvii. See also Herschel, *Preliminary Discourse*, p. 262: 'it is by no means impossible that the Newtonian theory of light, if cultivated with equal diligence […] might lead to an equally plausible explanation of phenomena now regarded as beyond its reach.'

depended on comparatively small-scale and accessible experimentation. Herschel thought it a mark of Young's genius that Young had made such extraordinary advances using such homely equipment:

A scrap of card, a hair, and a candle; a few scratches on a bit of glass held in the sun and turned slowly round; a piece of paper, a pinhole, and a closed window shutter, was all the apparatus he required, and proved abundantly sufficient to satisfy himself and every one else of the truth of a physical law so elegant as to command universal attention, and so important as at once to change the face of optical science.[36]

It is quite possible that Herschel, a founding member of the Analytical Society which had imported advanced Continental mathematical methods to English science, saw the rhetorical advantages of counterbalancing the association of the undulatory theory with foreign mathematics by emphasizing the simple, domestic nature of the optical equipment used by the wave theory's first and greatest British exponent. But it was certainly true that basic optical experiments could be performed using materials easily available in kitchens, drawing-rooms, and toyshops. The Victorian enthusiasm for optical illusions, toys, and demonstrations has been well documented by Carol T. Christ, Hal Foster, and others. Kate Flint rightly argues that

throughout the period, Victorians were fascinated with the technology of vision—whether the apparatus concerned was located outside the body (as in the case of the telescope, the microscope, the camera lens or optical toys) or whether it was an integral part of it—the eye.[37]

Underlying this practical and philosophical interest in optics was the complex of experimentation and mathematical analysis which resulted in the triumph of the wave theory of light at the onset of the Victorian era.

The result of the accessibility of optical equipment and the theological interest in light was that as the battle between the corpuscular and undulatory theories was being fought out at the highest levels of European science, a slew of quasi-scientific works claiming to present new views of the nature of light were being published in Britain by a wide variety of writers. These works are a very interesting case study in the

[36] J. F. W. Herschel, *Letter from J. F. W. Herschel, Esq.* (No place of publication, no publisher: 1830[?]), pp. 6–7.
[37] Kate Flint, *The Victorians and the Visual Imagination* (Cambridge: CUP, 2000), p. 311.

interrelations of science, religion, and literature in the early nineteenth century. Particularly interesting are those which are conscious of lying outside the structures of mainstream scientific practice or discourse, because these tend to defend their own legitimacy via various kinds of critiques of science.

These books often highlight eddies and whirlpools in the tide of professionalization and specialization in the sciences. They explicitly challenge the professional identity of establishment science by seeking to overturn emerging disciplinary boundaries, by failing to respect professional etiquette of various kinds, and by disregarding the caution and gradualism which were becoming hallmarks of scientific discourse. Frequently, their personal narratives suggest that the authors have a better claim to attention by virtue of their status as outsiders than the hidebound, tradition-blinded scientific establishment. Thus, these books both acknowledge and reject the developing standards and protocols of professional science. Indeed, the very fact that they were published as books indicates their outsider status. Their authors were neither participants in the beginnings of the system of peer-reviewed papers in journals nor synthesizers of established scientific doctrine like Mary Somerville or even Alexander von Humboldt, whose book-length publications for the general reader were supported by the writers' reputations within the scientific establishment.

A very interesting example is John Kyan's 1838 book *On the Elements of Light, and their Identity with those of Matter […]*. It argued that light was best understood via a wholly new conception of chemical elements. Kyan acknowledged only three chemical elements, compared with the fifty or so recognized by mainstream British chemistry at this time, and reclassified all other substances as compounds. This was because—in an idiosyncratic interpretation of the early nineteenth-century emphasis on the unity of nature—Kyan felt that a smaller number of elements would be 'in better accordance with true philosophy'.[38] Light, he argued, was a compound made up of oxygen, nitrogen, and hydrogen, and thus was a material substance. But Kyan did not support the corpuscular theory. Instead, he proposed a rethinking of ether science:

the theory of the existence of an ethereal substance pervading space, penetrating all material bodies, and occupying the interstices between their molecules,

---

[38] John Howard Kyan, *On the Elements of Light, and their Identity with those of Matter, Radiant and Fixed* (London: Longman *et al.*, 1838), p. 4.

I consider perfectly sustained, and [...] *light*, or its constituents, in their molecular state, is *that ethereal substance*.[39]

Kyan's proposal was just one of a large number of radical re-envisionings of contemporary chemical and physical science, many of which seem to have been driven by either religiously motivated antipathy to science altogether or by a desire to tidy and codify the confusing expansion of scientific principles into a smaller and more manageable number. The interest of Kyan's book lies in its author and his negotiation of his relationship with the scientific establishment, rather than in his chemical theory.

Of the many attempted interventions by non-specialists in the debate on the nature of light, Kyan's was perhaps the most conscious of the need to offer an alternative to the rhetoric of scientific authority. His book is elaborate in its attempts to please the reader. It was an expensive production, illustrated with lovely hand-painted colour wheels and spectra drawn by George Field and engraved by David Lucas. As well as these attractions, an urbane passage dedicates the book to 'the Ladies of the British Isles [...] who visit our Philosophical Associations and Institutions and, by their cheering presence, do so much for the encouragement and cultivation of the Sciences'.[40] Despite claiming to be grateful for the encouragement of the ladies of the nation, Kyan goes on to emphasize that he lacks the privileges of a scientific insider. Like many quasi-scientific texts by non-practitioners of science, the book opens with substantial prefatory material acknowledging the writer's lack of standing in the scientific world and the book's unorthodox argument. Announcing that the book will contain 'STARTLING ASSERTIONS', Kyan insists that contemporary science is rigidly conservative and restricted in its imagination and that it chains up 'the inspiring spirit of investigation'.[41] Nonetheless, he respects the scientific establishment enough to hope that his book will inspire 'the master-mind of some eminent Philosopher to enlist in the investigation [of light], which may be productive of incalculable benefit'.

The philosopher whom Kyan had chiefly in mind was Michael Faraday.[42] Nearly twenty years earlier Kyan had sent two hundred of 'the most eminent philosophers of the day' a pamphlet describing his system for reducing the number of chemical elements, and had waited

---

[39] Ibid., p. 40.   [40] Ibid., p. iv.   [41] Ibid., p. v.   [42] Ibid., p. 1.

to hear their responses. Not a single one, it appears, had replied.[43] By the late 1830s, however, Kyan had greatly strengthened his claim to the interest of the scientific world, and to that of Faraday in particular, by inventing a new and apparently highly effective method of preserving timber, an invention which made Kyan a great deal of money and gave him a national reputation.

In 1836 Parliament passed an act authorizing the raising of the immense sum of a quarter of a million pounds as the capital for the Anti-Dry Rot Company, to which Kyan sold his rights in the process. *Bentley's Miscellany* attested to the celebrity of Kyan and his invention, celebrating the Anti-Dry Rot Company with a fantasy about the Muses preserving their sacred trees by kyanizing them:

At the office in Lime-street-square the Muses bargained for a steeping of their undying, dying, decaying timber in the wondrous tank at Red Lion wharf, Poplar. The process, notwithstanding the mischief done to the wood by the poets of this scratching age, was most triumphantly successful; all symptoms of decay, except where certain initials were carved, at once disappeared, and the immortal plants at once began to put on 'all their original brightness!' [...] The Muses, with a few select friends, dined together afterwards at the Macclesfield Arms in the New Road [...].[44]

Kyan had a considerable national reputation, then, in 1830s chemical technology. His invention was of course of enormous possible application to both civil and naval construction, and this led to his first connection with Faraday, who chose kyanization as the topic of his 1833 inaugural lecture as Fullerian Professor of Chemistry at the Royal Institution, a body whose mission to make science useful Faraday took seriously.[45] Over the next few years, Faraday was called on by the Admiralty and by Isambard Kingdom Brunel to give expert opinions about the efficacy of Kyan's preservation process for ship and railway building.[46] It was perhaps natural therefore for Kyan to look to Faraday for support when he sought to capitalize

[43] Kyan, *On the Elements of Light*, pp. 4–5.

[44] Anon., 'Kyan's Patent—The Nine Muses,—and the Dry-Rot', *Bentley's Miscellany*, 1 (1837), 93–6 (p. 94).

[45] For discussion of Faraday's religious understanding of the application of science to human purposes, see Geoffrey Cantor, *Michael Faraday: Sandemanian and Scientist* (Houndmills: Macmillan, 1991) pp. 194–5.

[46] See Frank A. J. L. James, ed., *The Correspondence of Michael Faraday* (London: Institution of Electrical Engineers, 1991–), II, pp. 264, 301, 425 and 432–3 for letters discussing Faraday's work with the 'kyanising' process.

on his reputation in chemical technology to support his work on the fundamental theory of chemistry; but his efforts failed. Faraday did not, as Kyan had hoped he would, support his work on light, and his 'startling assertions' went unanswered for a second time. In Kyan's case there was too great a gulf between industrial technological expertise and the kind of reputation needed to back theoretical scientific work; the one could not be translated or finessed into the other, no matter how lavishly illustrated or how urbanely dedicated the attempt.

Kyan's theory tried to hybridize the basic assumptions of the wave and corpuscular theories by explaining light as both material and ethereal. This hybridity made his work unusual among the attempted interventions by non-specialists in the debate over the nature of light. More often, such writers stuck dogmatically to the corpuscular side of the argument. It is striking that the overwhelming majority of non-scientists seeking to contribute to the debate weighed in on this side. A large number of these anti-undulationist interventions were made: they demonstrate the interest the topic roused far outside the scientific establishment. Few of these books are so handsome a production as Kyan's, which was presumably backed by the money he made from selling his rights to the Anti-Dry Rot Company. And few of these writers are so interested in their relationship with the scientific establishment. But the books do give a good sample of the way in which non-scientists weighed in on the pro-corpuscular side of the debate about the nature of light. In two of the following examples, the writers were quite open about the fact that the reasons for their antipathy towards the wave theory were religious rather than scientific.

Charles Bompass, a London barrister, published his *Essay on the Nature of Heat, Light, and Electricity* in 1817. Bompass assumed a material explanation of all three phenomena, including light: 'that light is matter, is so evident, and so generally admitted, that it is not necessary here to attempt to prove it'.[47] He also supported material explanations for other phenomena, including the theory of caloric, which saw heat as a fluid, and the 'two-fluid' theory of electricity. In all three cases Bompass was relying on outdated information. Almost two decades earlier, Rumford and Davy (among others) had done for heat and electricity what Young had done for light: that is, they had rejected

---

[47] Charles Carpenter Bompass, *An Essay on the Nature of Heat, Light, and Electricity* (London: Underwood, 1817), p. 151.

'imponderable fluid' explanations.[48] Though by no means universally accepted, Rumford's account of heat as the vibrations of matter and ether, and Davy's view that electricity was the result of chemical affinity and not the activity of a fluid or fluids, were major parts of the scientific debate in a way that Bompass did not recognize.

John G. Macvicar's 1833 *Inquiries Concerning the Medium of Light and the Form of its Molecules*, like Bompass's book, argued that light was made up of material particles, but went further than the barrister, hypothesizing a single unified theory which would demonstrate that 'heat, gravitation, affinity, magnetism, electricity and electro-magnetism may be explained on mechanical principles familiar to all and disputed by none'.[49] Macvicar had spent a short time as a lecturer in natural philosophy at St Andrews but devoted the rest of his career to the Presbyterian church, first as a missionary in Ceylon and then in a parish in Dumfriesshire, though he continued to publish on science as well as on philosophical and religious subjects. Macvicar's argument about light was driven by his Christian and anti-materialist principles. In order to assert that only the 'Creating and Sustaining Power' of God could really explain natural phenomena, Macvicar found it necessary to identify all those phenomena as spiritually inert, and hence material. Light could not be allowed to be immaterial because that would risk imbuing it with some kind of 'spontaneity or intelligence', which might detract from God's uniqueness.[50]

Another quasi-scientific outsider, W. M. Higgins, writing a few years earlier, was equally hostile to the idea that light was immaterial. He expressed sarcastic surprise that 'notwithstanding the ease and elegance which characterize all attempts which are made to account for Optical phenomena upon [material] principles, there have been and are men, of the greatest talent and learning, who have supposed light to be an immaterial essence'.[51] One of Higgins's most keenly urged objections to the immateriality of light was the fact that material objects such as mirrors, prisms, and crystals could cause light to be reflected and refracted, which seemed below the dignity of an immaterial substance.

---

[48] See Harman, *Energy, Force, and Matter*, p. 19.

[49] John G. Macvicar, *Inquiries Concerning the Medium of Light and the Form of its Molecules* (Edinburgh: Black; London: Longman *et al.*, 1833), p. 132.

[50] Ibid., p. 132.

[51] W. M. Higgins, *An Introductory Treatise on the Nature and Properties of Light, and on Optical Instruments* (London: Nimmo; Edinburgh: Blackwood; Dublin: Vurry, 1829), p. 15.

'It cannot be supposed that a material lifeless mass of matter can act upon an active and immaterial agent'.[52] Higgins imposed a scheme of differential moral values on the matter and forces in the universe and sought to subordinate the former to the latter.

In their different ways, all these attempted interventions from non-specialist writers were seeking to establish the materiality of light. Their consciousness that the main thrust of contemporary science was against them varied; for the most part they seem to have seen their opponents as materialists rather than as specifically wave-theorists. It might seem paradoxical that in order to fight potentially anti-religious materialism they were obliged to insist on light's materiality: but they were committed to maintaining a strict separation of matter and spirit or mind, and an immaterial light which sometimes behaved in material ways (for instance by affecting chemical change in substances exposed to it) threatened to undermine that separation.

On the other side of the debate, though, wave theorists might indeed consider light immaterial, but there was no general agreement among them about how material was the ether in which light waves moved. William Whewell, as we have seen, was prepared to argue that the ether was material enough to slow down a planet's orbit, but the more cautious Baden Powell was 'content to look at the theory simply as a mathematical system which faithfully represents, at least, a wide range of phenomena', since 'we certainly have no independent evidence' of the 'actual existence of a material fluid medium' for light.[53] But supporters of the wave theory often used the more straightforwardly material nature of light as explained by the corpuscular theory to attack that theory. A key instance, which appears in many pro-undulationist documents, particularly those intended for a non-specialist readership, was Herschel's calculation that even a minute particle, if it were travelling at the known speed of light, would arrive at the eye with the force of a 150lb cannon ball.[54]

On both sides of the debate about the nature of light, then, the topic of materiality was of crucial importance. To anti-materialists like W. M. Higgins, the suggestion that light was immaterial blurred the boundaries between mind and matter to a degree that was unacceptable

---

[52] Ibid., p. 16.
[53] Powell, *A General and Elementary View of the Undulatory Theory*, p. iv.
[54] J. F. W. Herschel, 'Light', *Encyclopaedia Metropolitana*, 29 vols (London: Fellowes, 1830), IV, pp. 341–586.

for philosophical and religious reasons. But most undulationists would have strongly resisted the idea that the wave theory implied a materialism threatening to religion. Of the central wave theorists, Whewell adopted the most natural theological approach, arguing that the nature of light was an indication of God's planning for the good of humanity: 'as the eye is made for light, so light must have been made, at least among other ends, for the eye'.[55] As for the ether, its materiality was similarly to be understood as a manifestation of divine design: 'the mere material elements [...] are full of properties which we can understand no otherwise, than as the results of a refined contrivance'.[56] Though Whewell was writing here in a context particularly designed to highlight the happy blending of science and religion, his Christian-inflected approach to questions of the moral value of matter and materiality was not untypical of that of the 'gentlemen of science' who dominated most of the British scientific establishment in the early nineteenth century.

FARADAY AND THE BEGINNINGS OF FIELD THEORY

All that is solid melts into air [...][57]

In his late twenties, Faraday was a member of the City Philosophical Society, which met fortnightly in Dorset Street, London, providing scientific discussion evenings, and carefully avoiding political subjects so as not to fall foul of the Seditious Meetings Act.[58] In 1819 he noted down a question about the transmission of forces that he intended to raise at the C. P. S.:

"Bodies do not act where they are not." Query is not the reverse of this true? Do not all bodies act where they are not, and do any of them act where they are?[59]

---

[55] Whewell, *Astronomy and General Physics*, pp. 128–9.   [56] Ibid., p. 137.
[57] Karl Marx and Friedrich Engels, *The Communist Manifesto* (1848), trans. Samuel Moore (1888), ed. Gareth Stedman Jones (London: Penguin, 2002), p. 223.
[58] James Hamilton, *A Life of Discovery: Michael Faraday, Giant of the Scientific Revolution* (New York: Random House, 2002), p. 129.
[59] Faraday, 'Common Place Book', Archive of the Institution of Electrical Engineers, Special Collections MS2, p. 324.

Claiming that bodies could only act 'where they are not', Faraday was envisaging space as a kind of neutral container within which matter forms solid and impermeable bodies. That is to say, he was thinking with the mechanical, atomist view of space and matter which had dominated British physical science since Newton. This view made a sharp distinction between space and the bodies—even at the tiny scale of atoms—that exist in it. Already this distinction had been called into question by Davy and others.[60] But the development of field theory from the mid-1840s on dramatically changed the nature of the debate. In 1846, in a seminal paper called 'Thoughts on Ray-vibrations', Faraday showed how far he had moved from his early speculations about the ways in which forces moved between objects in space. Cautiously framing his proposals with disclaimers about their resulting from 'the vague impressions of my mind', and 'the shadow of a speculation', Faraday nonetheless made a series of radical suggestions about the nature of space and matter.[61] The result of these suggestions was to abolish the distinction between bodies and the space they occupy.

Since at least 1844 Faraday had regarded atoms not in the Newtonian manner as tiny bodies surrounded by forces, but instead as 'centres of force'. Under this view atoms did not have solid bodies at all but consisted entirely of intensifications in the forces which pervaded all space.[62] In his 'Ray-vibrations' paper he built on this view, extending the idea that atoms were centres of force to argue that both space and

[60] David Knight notes that even in the first decade of the nineteenth century, Davy's atomic theory was influenced by non-Newtonian views: *Humphry Davy, Science and Power* (Cambridge: CUP, 1992), p. 76. Pearce Williams argues that 'from 1815 until the end of his life, Davy thought of matter in terms of point atoms surrounded by attractive and repulsive forces' (*Michael Faraday: A Biography* (London: Chapman and Hall, 1965), p. 79).

[61] Faraday, 'Thoughts on Ray-vibrations', in *Experimental Researches in Chemistry and Physics* (London: Taylor and Francis, 1859), pp. 366–72 (pp. 367, 372).

[62] There has been considerable debate about the exact extent to which Faraday's atomic theory at various stages of his career resembles the eighteenth-century Jesuit scientist Roger Joseph Boscovich's 'point atom' theory. L. Pearce Williams argues that Faraday's adoption of Boscovich's theory should be recognized as underpinning claims that his science had a theoretical foundation 'of any importance' (*Michael Faraday*, p. 78; and see also pp. 129–30); but the investigation of Faraday's specific indebtedness to Boscovich is made less urgent by the fact that—as Geoffrey Cantor explains—Faraday was far from the only contemporary British scientist to adopt and adapt Boscovichean theory, and in the light of the fact that, as points out, Nancy J. Nersessian 'Faraday, himself, probably never read Boscovich' (Cantor, *Michael Faraday*, p. 175; Nersessian, *Faraday to Einstein: Constructing Meaning in Scientific Theories* (Dordrecht: Martinus Nijhoff, 1984), p. 39).

matter were made up of greater or lesser intensities of forces. In a sort of helpless climax he wrote: 'all I can say is, that I do not perceive in any part of space, whether (to use the common phrase) vacant or filled with matter, anything but forces and the lines in which they are exerted'.[63] In 'the common phrase'—which really meant the common idea—there were such things as empty space and solid matter. But the purpose of Faraday's argument was fundamentally to overturn this belief.

Faraday contrasts the quasi-Newtonian understanding of bodies with his own developing one. The traditional view sees objects as having 'a definite form and a certain limited size', but in Faraday's new theory, 'that which represents size may be considered as extending to any distance to which the lines of force of the particle extend; the particle indeed is supposed to exist only by these forces, and where they are it is'.[64] The physical boundaries of a body, then, are not those perceptible to the eye or the touch, but instead they are the limits of the forces the bodies exert. For example, since a magnet exerts forces on iron filings even at the distance of (say) several inches, the magnet must be thought of as extending as far as those filings, even though we perceive it as a bounded object some distance from them. The magnet, indeed, *is* the forces it exerts. And not only magnetic but also electrical, gravitational, and probably other forces too act in this same way.[65] In his new conception, then, it would not be enough to say that forces pervade all space and all the bodies in it; instead, space and matter alike *consist in* these forces.

Faraday's key innovation here was his conception of lines of force. Since the early 1830s he had been investigating the idea that magnetism and electricity travelled through space in curved lines. Iron filings sprinkled around a magnet form into patterns of lines that bell out from one of the magnet's poles and then back towards the other. Where these lines are most closely clustered together, the strength of the magnetic force is greatest. Faraday came to see these lines not as a way of imagining behaviour of forces but as physical, as having real existence. Faraday developed the view that—as Pearce Williams puts it—each line of force 'was an actual entity, somehow associated with matter, but also independent of it'. Pearce Williams adds: 'it is no exaggeration to say that a fundamentally new way of looking at physical reality was introduced into science' in Faraday's experiments on the lines

---

[63] Faraday, 'Thoughts on Ray-vibrations', p. 370.
[64] Ibid., p. 367.   [65] Ibid., p. 369.

of force.[66] For our purposes, the most important thing to be noted is that traditional beliefs about matter having extension and solidity were replaced in Faraday's work with a concept of physical but not exactly material lines of force. This concept necessitated a redefinition not only of matter and the material but also of space.

If solidity and extension were no longer real properties of matter, but only illusions created by limited human sense mechanisms, equally, space was no longer either full or empty in the traditional sense. At the end of his paper 'On Electric Conduction and the Nature of Matter', Faraday described his new theory as resulting in a vision of space as completely full, but full of force rather than of solid objects: 'matter fills all space, or, at least, all space to which gravitation extends [...] for gravitation is a property of matter dependent on a certain force, and it is this force which constitutes the matter'. This meant that 'matter is not merely mutually penetrable, but each atom extends, so to say, throughout the whole of the solar system, yet always retaining its own centre of force'.[67] Geoffrey Cantor argues that this new vision of a universe filled with force appealed to Faraday on religious grounds. Empty space threatened to disrupt Faraday's 'metascientific principle of plenitude', but if space can only exist where force is present, emptiness becomes a logical impossibility and a new version of the plenum comes into being.[68]

This new kind of fullness did away with the idea of solidity as it had been understood; and it also did away with the ether. The lines of force could explain everything that ether explained about how electricity and other forces were transmitted through space, and they could do it without supposing the contradictory physical properties with which ether had to be endowed.[69] Faraday was profoundly dissatisfied with the compromises of known physical laws that were necessary to make ether explicable. Instead, stripping ether to its fundamental properties of being made up of almost infinitely small nucleii and having almost infinitely intense elasticity, Faraday resolves it into his universe-pervading forces: 'if such

---

[66] Pearce Williams, *Michael Faraday*, pp. 204, 205–6.

[67] Faraday, 'On Electric Conduction and the Nature of Matter' (1844), in *Experimental Researches in Electricity*, 3 vols (London: Taylor and Francis, 1839–55), (II, p. 293).

[68] Cantor, *Michael Faraday*, p. 182. But Olivier Darrigol argues that Faraday's position was that the universe was made up of 'gravitational, electric and magnetic lines of force crossing empty space' (*Electrodynamics from Ampère to Einstein* (Oxford: Oxford University Press, 2000), p. 112).

[69] Faraday, 'Thoughts on Ray-vibrations', pp. 370–1.

be the received notion [of ether], what then is left in the ether but force or centres of force?'.[70] Faraday's conception was more compelling than the ether model, judged by the early and mid-nineteenth-century sense of the unity of nature, because while ether could only be part of a dyad with matter, Faraday's forces were in themselves both space and matter.

Faraday's rejection of ether did not last. His work was developed into 'classic' field theory by William Thomson and James Clerk Maxwell, who made it expressible in the language of advanced mathematics, and reintroduced a unified ether.[71] Maxwell's crucial assertion in 'On Physical Lines of Force' that 'we can scarcely avoid the inference that *light consists in the transverse undulations of the same medium which is the cause of electric and magnetic phenomena*' brought Faraday's field theory, ether, and the wave theory of light together in a triumphant statement of the unity of natural forces.[72] But even though ether was back, neither matter nor space could be understood separately any longer.

Neither the Victorian scientific establishment, nor contemporary writers in other fields, were quick to take up the implications of Faraday's speculations on the nature of matter and space. Literary criticism, certainly, has been much more interested in field theory as it developed after Einstein than in 'classic' field theory. Tim Armstrong interestingly suggests that 'one result of field theory is a heightened sense of intertextual relations', but he makes this suggestion in the context of an argument about twentieth-century American poetics.[73] As far as I

---

[70] Faraday, 'Thoughts on Ray-vibrations', p. 369.

[71] A decade after 'Thoughts on Ray-vibrations', the 25-year-old Maxwell gave a 'career-making' paper 'On Faraday's Lines of Force' to a gathering of elite Cambridge mathematicians in 1856, and in the early 1860s published another, 'On Physical Lines of Force' in instalments in the *Philosophical Magazine* (Crosbie Smith, 'Force, Energy and Thermodynamics', in *The Cambridge History of Science*, vol. 5, *The Modern Physical and Mathematical Sciences*, ed. Mary Jo Nye (Cambridge: CUP, 2003), pp. 289–310 (p. 305)). The first paper considered the geometrical aspects of the lines of force; the second emphasized their physicality and generated a new kind of ether theory, in which the same medium that carried electrical and magnetic forces also propagated light. A very helpful account of Maxwell's responses to Faraday and to Thomson (the other great contributor to early field theory based on Faraday's investigations) is given in P. M. Harman's *The Natural Philosophy of James Clerk Maxwell* (Cambridge: CUP, 1998), pp. 71–90.

[72] James Clerk Maxwell, 'On Physical Lines of Force', in *The Scientific Papers of James Clerk Maxwell*, ed. W. D. Niven, 2 vols (Cambridge: CUP, 1890; rep. Mineola, N.Y.: Dover, 2003), I, 451–513 (p. 500).

[73] Tim Armstrong, 'Poetry and Science', in Neil Roberts, ed., *A Companion to Twentieth-Century Poetry* (Oxford: Blackwell, 2001), pp. 76–88 (p. 84). Armstrong cites especially Charles Olson's essay 'Equal, That Is, To the Real Itself' (1958) as re-making

know there have been no major attempts to read Victorian intertexuality in the light of field theory.

But there has been at least one sustained and successful literary critical engagement with early field theory, though not in the area of intertextuality. Daniel Brown's study of Hopkins explores the implications of the fact that an education at Oxford and in the Jesuit novitiate exposed Hopkins to contemporary physics. Brown reads Hopkins's vocabulary of 'stress' and 'strain' in the context of the theories of a 'solid elastic ether' which developed in Maxwell's work on Faraday's field theory.[74]

Brown argues that the idea of an ether in which various kinds of vibration and tension caused different physical phenomena—such as light, magnetism and electricity—led to a conception of a physical universe constantly under an immense number of stresses and strains. In particular he shows how Maxwell's use of 'stress' to replace the Newtonian ideas of action and reaction made that word a key part of the vocabulary of physical science in the 1850s and 60s.[75] Brown goes so far as to speculate that Hopkins's key neologism 'instress' may have been a contraction of 'internal stress', a phrase used by the Scottish physicist William Rankine in an 1858 argument about the distribution of forces within bodies.[76] More generally, he argues that 'nature is often depicted by Hopkins in a suggestive analogy to the dynamic principle of the physical field as a plenum of activity'.[77] That is, he sees Hopkins's tendency to fill all the spaces of his natural scenes with what we might think of as events (including sounds, sights, feelings, and associations as well as activities) as at least 'akin' and perhaps directly related to the ideas of a universal ethereal medium being developed by the field theorists of his generation. Brown's ambitious and compelling account of Hopkins's imagining of an ether shows Hopkins aligning the medium that pervades universal space with God's love and with the intercession of the saints. His reading of 'The Blessed Virgin Mary compared to the Air we Breathe', especially, interprets Mary as 'a medium of propagation both in the bodily sense, as a mother, and in the manner of the ether in field theory'.[78]

Following Brown's lead we might well be stimulated to investigate other scientifically educated late nineteenth-century writers for entities

---

field theory to support a vision of reality as 'a dynamic arrangement of forces or pathways' (p. 84). The Olson essay is in *Collected Prose*, ed. Donald Allen and Benjamin Friedlander (Berkeley: University of California Press, 1997), pp. 120–5.

[74] Daniel Brown, *Hopkins' Idealism: Philosophy, Physics, Poetry* (Oxford: Clarendon Press, 1997), p. 212.

[75] Ibid., pp. 213–18. [76] Ibid., p. 219. [77] Ibid., p. 241.

[78] Ibid., pp. 244–45.

that can be read as analogies of ethereal media. For example, a number of recent critics have drawn interesting connections between spiritualism in popular literature and practice and late Victorian ether theories, thereby leading the analysis of the cultural impact of field theory into areas to which Faraday would have vehemently objected.[79]

If we simplify the variety of mid-nineteenth-century field theories to their central shared proposition—if we consider them as ways of exploring the effects that bodies in space have on other bodies around them—we may find another useful means of describing interrelations between elements in the complex systems of novels, for instance. Eliot's famous 'web' metaphor in *Middlemarch* has of course been a touchstone for a number of important and influential studies of the novel which read it in the context of contemporary evolutionary theory.[80] But to read the society of Middlemarch as a field made up of lines of force (produced perhaps by wealth and sexual desirability, among other things) allows us to envisage the characters and the town as consisting in their various relations (kinship, patronage, romantic love, and so on) in the same way that objects are made up by forces in classic field theory. As well as an

---

[79] Faraday's extended battle against the invocation of electromagnetic and other scientific explanations by believers in spiritualism is discussed briefly in Hamilton, *A Life of Discovery*, pp. 352–4. In a letter on this subject to *The Times* (28 June 1853), which he thought important enough to reprint in *Experimental Researches in Chemistry and Physics*, Faraday maintained a courteous tone until the last few lines, where he shows his irritation with those proponents of spiritualist beliefs who 'reject all consideration of the equality of cause and effect, who refer the results to electricity and magnetism, yet know nothing of the laws of these forces [...] or who even refer them to diabolical or supernatural agency, rather than suspend their judgment, or acknowledge to themselves that they are not learned enough in these matters to decide on the nature of the action' ('On Table-turning', in *Experimental Researches in Chemistry and Physics*, pp. 382–5 (p. 385). For examples of nineteenth-century writers suggesting that field or ether theories could give a scientific explanation for spiritualist phenomena, see Richard Noakes, 'Spiritualism, Science and the Supernatural in Mid-Victorian Britain', in Nicola Bown, Caroline Burdett and Pamela Thurschwell, eds, *The Victorian Supernatural* (Cambridge: CUP, 2004), pp. 23–43, esp. pp. 29–30), Roger Luckhurst, *The Invention of Telepathy* (Oxford: Oxford University Press, 2002), esp. p. 88, and Janet Oppenheim, *The Other World: Spiritualism and Psychical Research in England, 1850–1914* (Cambridge: CUP, 1985), esp. pp. 379–87; see also Steven Connor's very stimulating account of the emphasis in Victorian spiritualism on explaining apparent action at a distance in 'Voice, Technology and the Victorian Ear', in *Transactions and Encounters: Science and Culture in the Nineteenth Century*, ed. Roger Luckhurst and Josephine McDonagh (Manchester: Manchester University Press, 2002), pp. 16–29, esp. 23–5.

[80] The key example, of course, is Gillian Beer's discussion in *Darwin's Plots: Evolutionary Narrative in Darwin, George Eliot and Nineteenth-Century Fiction* (London: Routledge, 1983; 2nd edn, Cambridge: CUP, 2000) esp. pp. 139–75.

organic web we can see Middlemarch as a field in which people—to adapt Faraday's argument in 'Thoughts on Ray-vibrations'—are not bounded, self-contained units operating in a neutral space but extend as far as their influence does, creating the space of the town as they move within it. There is no reason why the two kinds of scientific metaphor should not co-exist in the same text. Eliot herself brings together organic science and atomic physics in her description of Lydgate's contempt for the products of a shoddy literary or artistic imagination compared with his admiration for the work of the scientific imagination:

these kinds of inspiration Lydgate regarded as rather vulgar and vinous compared with the imagination that reveals subtle actions inaccessible by any sort of lens, but tracked in that outer darkness through long pathways of necessary sequence by the inward light which is the last refinement of Energy, capable of bathing even the ethereal atoms in its ideally illuminated space.[81]

The 'subtle actions' in which Lydgate is most interested are those of the human body and the bodies that invade it, but the 'inward light which is the last refinement of Energy' is a non-corpuscular light, a product of the triumph of the wave theory, and the 'ethereal atoms' belong to the debate—running in the decade in which *Middlemarch* is set as well as in that in which it was written—about the fundamental nature of matter.[82]

It is not at all surprising that Eliot should be interested in the new ideas about matter; neither is it surprising that the Victorian novel should be deeply concerned with ways of representing the material as an index or even a product of character. Ralph Nickleby surrounds himself with the solidity of material objects in order to manifest certain aspects of his character, or what he hopes others will take to be his character. The narrator in *Middlemarch* uses the solidity of objects as an example of knowledge that is so deeply ingrained that it cannot be questioned, and links this knowledge to the humane task of mutual respect:

We are all of us born in moral stupidity, taking the world as an udder to feed our supreme selves: Dorothea had early begun to emerge from that stupidity,

---

[81] George Eliot, *Middlemarch* (1871–2), ed. David Carroll (Oxford: Clarendon Press, 1986), p. 162 (ch. 16).

[82] Eliot's irony here is that while Lydgate is glorifying his interest in the deep questions of nature, he is failing to consider his relationship with Rosamund, which will ultimately put a stop to all his scientific ambitions. He ought to be paying attention to the 'subtle actions' of Rosamund's wiles, which are 'inaccessible' to his lenses but are 'tracked [...] through long pathways' by Eliot's narrative examination.

but yet it had been easier to her to imagine how she would devote herself to Mr Casaubon, and become wise and strong in his strength and wisdom, than to conceive with that distinctness which is no longer reflection but feeling—an idea wrought back to the directness of sense, like the solidity of objects—that he had an equivalent centre of self, whence the lights and shadows must always fall with a certain difference.[83]

Realism's commitment to the investigation of character via the material means that it has sometimes to explore the immateriality of matter. By 1870 it was clear that the 'solidity of objects' is partly illusory. Samuel Johnson's attempt to refute Berkeley's anti-materialism by 'striking his foot with mighty force against a large stone' is not enough to persuade us that what Eliot calls the directness of sense guarantees a materialist view of the world.[84] In *Middlemarch*, kicking stones does not tell us much, though the stones themselves may be full of information:

the stone which has been kicked by generations of clowns may come by curious little links of effect under the eyes of a scholar, through whose labours it may at last fix the date of invasions and unlock religions [...].[85]

Realism is characterized by a similar effort to open history out of unregarded things, and Eliot's realism is strengthened by her acknowledgement that the 'directness of sense' may conceal as much as it reveals. The solidity of material objects in Eliot's great novels is a comforting illusion, a way of living with the self rather than of understanding the world. When Tom comes home from school early in *The Mill on the Floss* he is warmed by familiar surroundings: 'the pattern of the rug and the grate and the fire-irons were 'first ideas' that it was no more possible to criticize than the solidity and extension of matter'.[86] But the solidity and extension of matter *are* open to criticism by Eliot, if not by Tom; and by linking them with childhood perceptions Eliot places them in the past at the same time as she invests them with nostalgic affection. The passage continues:

There is no sense of ease like the ease we felt in those scenes where we were born, where objects became dear to us before we had known the labour of choice, and where the outer world seemed only an extension of our own personality:

---

[83] Eliot, *Middlemarch*, p. 205 (ch. 21).
[84] James Boswell, *Life of Johnson*, ed. R. W. Chapman (Oxford: Oxford University Press, 1980, repr. 1998), p. 333 (6 August 1763).
[85] Eliot, *Middlemarch*, p. 402 (ch. 41).
[86] George Eliot, *The Mill on the Floss*, (1860), ed. Gordon S. Haight (Oxford: Clarendon Press, 1980), p. 133 (book 2, ch. 1)

we accepted and loved it as we accepted our own sense of existence and our own limbs.

Eliot transfers 'extension' from matter to personality and links it to an immature, innocent egocentricity; and her implied repudiation of this childish perception of extension retrospectively distances her also from the extension of matter.

The nature of matter is profoundly open to question in *Middlemarch* and *Mill on the Floss*, and calling into doubt the reliability of matter helps Eliot to investigate the field of connections and mutual influences which is her topic in these novels. I think it is not too much to suggest that *Middlemarch* is the first great artistic expression of a sensibility influenced by field theory and its understanding of each body (each person) as 'not merely mutually penetrable, but [extending], so to say, throughout the whole of the solar system, yet always retaining its own centre of force'.[87] Of course *Middlemarch* is a key locus for a great many possible investigations into nineteenth-century literary and scientific interactions. But my argument here is not that it is because *Middlemarch* is interested in science that it is an important text for enquiring into literary responses to field theory. Rather, I want to suggest that *Middlemarch*'s importance in this enquiry is due to its representing a certain kind of realism, one which is centrally concerned with methods of connection and means of transmission; and therefore that other texts with these concerns may also be suitable for investigation in the light of field theory, though their authors lacked Eliot's scientific knowledge. Field theory obliges us to be sceptical about beliefs in limits, in solidity, and in neutral space. In terms of fiction these are the beliefs that give rise to ideas about the self as a bordered, atomic unit; about the boundaries of the influence exerted by people on one another; and about the neutrality of the spaces within which they live. These beliefs appear to be fundamental to the realist novel, and for that reason they are of course central concerns for many of those novels; and field theory, just as much as evolutionary theory, can help us to explore how Victorians understood those concerns.

[87] Faraday, 'On Electric Conduction and the Nature of Matter', p. 293.

7

Chaos, the Void, and Poetry

Previous chapters have explored ways in which early nineteenth-century writers valorized space as a fundamental of science, as a transcendental abstraction, and as a unified, homogeneous field which underlies the fragmentation of disciplinary formation. This chapter shifts the focus from space as reliable, ordered, and ideal to space as random, unstable, and chaotic. With this shift we move from science to pseudo-science, and from non-fiction prose to poetry—albeit a kind of poetry that has never had a place in the critical canon.

Chaos seems to have had a particularly powerful place in the imaginations of early nineteenth-century writers, and it is not difficult to construct a set of reasons why, in a time of acute ideological and social conflict, this should be so. But instead of looking at political representations of chaos in the period, which generally rely on inspiring a fear of disorder, this chapter focuses on texts which try to inhabit chaos imaginatively, making it a space of experimentation within a universe in which order is overwhelmingly dominant but sometimes of ambiguous value.

Along with chaos goes the concept of the void. Chaos is a hole in the fabric of the meaningful cosmos, a space in which distinctions, categorizations, and identities are emptied of significance. My last chapter's argument about the tendency towards dematerialization in early nineteenth-century physics and other disciplines needs an addendum: how did dematerialization feature in the literary culture of the period? The question is much too large to answer in a single chapter: instead I hope to give a partial response by examining the role of empty and disordered space in just one literary genre, one which was peculiar to the early nineteenth century: the cosmological verse epic. First, though, we have to return to the great model of this kind of poem: *Paradise Lost*.

MILTON, VOIDS AND THE EARLY NINETEENTH CENTURY

The void can be thought of as pure space: space as and for itself, not constituted as a container for or interpretation of nonspatial knowledge or experience. This purity makes the void highly problematic, and difficult to encompass within any ordered system, whether textual, theological, or philosophical. Equally, though, the void is the sign of the existence of an ordered system. Gérard Genette describes the plain style, with its marked absence of rhetorical figures, as a '*zero degree*, that is to say, a sign defined by the absence of sign'.[1] The zero degree 'is the infallible mark of the existence of a system'. A sign defined by absence can only be recognized, let alone interpreted, where a system exists.[2] A void space is a zero degree in this sense; we are only in a position to identify a gap or a hole when we have come to recognize the fabric in which the gap exists, and recognizing the existence of that fabric puts us on the road to recognizing the systems that construct it. The void as both a subject-matter and a constituent element of writing is evidence of the systems which underpin that writing. But nonetheless, a void may pose a challenge to the system it interrupts.

Voids shrug off verbal and cartographic attempts to describe them, and they draw the mind and the perceptions into logical impossibilities. They oscillate between states of being and nothingness, like a Gestalt figure where the white and black areas exist at the same time in the same frame to give two quite different pictures. Voids signify both absence and presence, and they can signify these at the same time. Tropes of absence and presence are embedded in thinking about systems of ordering and organizing—of systematizing, in other words—but we tend to think of them as mutually exclusive. The void, on the other hand, is the sign of their potential for conjunction.

In the Greek tradition, the world was created out of pre-existing matter rather than out of nothing at all. Empty space was an unsettling idea, which threatened the stability and logic of the world, but one which necessarily came into being along with the concept of space: as Rosalie

---

[1] Gérard Genette, *Figures of Literary Discourse*, trans. Alan Sheridan (Oxford: Blackwell, 1982), p. 47.

[2] Ibid., p. 48.

L. Colie points out, 'in obedience to dialectical symmetry, "nothing", "not-being", the void, and the vacuum all had to be considered as soon as "all", the cosmos, and the universe were conceived'.[3] The traditional *horror vacui* was reinforced in the physical sciences by Aristotle, who in Book 4 of the *Physics* argued that the void was a logical absurdity. For centuries of the Christian era, belief in the possibility of void space was widely held to be potentially atheistic, since space empty of matter implied space empty of God, and God could not be considered to be present in only part of the universe.[4] The long tradition of holding the possibility of the void anathema is felt in tensions in *Paradise Lost*. By the early nineteenth century, however, this theological tradition was far weaker and the idea of emptiness was ready for imaginative reappraisal, but of course Milton's poem was still dominant as a creative model and critical standard for cosmological poetry.

Paradise Lost introduced into the English language 'the void', a definable cosmological space distinct from 'voids'. The OED records a good deal of activity in the meanings of the term 'void' between the early seventeenth and the mid-eighteenth century, with new senses becoming more absolute and unequivocal. In the physical sciences, the possibility of actual void space was a subject of much controversy. Some twenty years before *Paradise Lost* was published, the first successful attempt to create a void space, or vacuum, was reported: an Italian experimenter, Evangelista Torricelli, inverted a test tube of mercury into a shallow pan containing the same liquid, and noted that some of the mercury in the tube slowly bled into the pan, leaving a gap at the top of the inverted test tube. He proposed that this gap was a truly void space, a vacuum.[5] Much controversy resulted from the reports of his discovery. Was the empty space a hole in the divinely ordered nature of the cosmos? In the light of contemporary debate regarding the theology of void space, Milton was taking a risk by giving as much importance as he did to the

[3] Rosalie L. Colie, *Paradoxia Epidemica: The Renaissance Tradition of Paradox* (Princeton, N.J.: Princeton University Press, 1966), pp. 220–1.

[4] Colie regards Pascal as the great exception to, and breaker of, this belief: 'for theologians before Pascal, the existence of a void space had seemed to threaten notions of providence, divine creativity, and benevolence. For Pascal, such a view was ludicrous': p. 258; see also p. 222 for another statement of the atheistic nature of post-Aristotelian belief in void space.

[5] See Steven Shapin and Simon Schaffer, *Leviathan and the Air-Pump: Hobbes, Boyle, and the Experimental Life* (Princeton, N.J.: Princeton University Press, 1985), p. 41.

void in *Paradise Lost*. He is careful to make his theological conception of space clear. God announces

> I am who fill
> Infinitude, nor vacuous the space.[6]

Since God has chosen not to use his power in the realm of Chaos, however, void space can exist there; and in associating Chaos with the void, Milton set up one of the key areas for investigation and confusion in early nineteenth-century cosmological poetry.

The vacua in Milton's Chaos are physical and literal, as well as metaphysical and metaphorical: and this is why Satan experiences turbulence during his flight through the abyss. Flying through Chaos he

> meets
> A vast vacuitie: all unawares
> Fluttring his pennons vain plumb down he drops
> Ten thousand fadom deep[7]

But for the most part, Milton's Chaos is not void in the sense of empty of matter. It is, however, void of order, of meaning. From the outside it appears to be empty: from within, it offers a surfeit of categories, too many and too unstable to constitute a comprehensible order. It is only while Satan is preparing for the journey that the space between Hell and earth is termed a 'void'; once he is inside this space it is named in terms of its multiple characteristics.[8] The elements of its landscape are enumerated as Satan travels 'o'er bog or steep, through strait, rough, dense or rare'; and Satan adds to these geographical terms a number of political ones, some of them mutually contradictory, including 'Realm', 'Empire', and 'Dominion'.[9] Once entered, Chaos becomes a place where names can be tried out but do not properly stick.

The Greek root of the word 'chaos' signifies a huge chasm or a void, and, according to the OED, the word continued to have this sense of a gulf or a void until 1667. But in the late sixteenth century 'chaos' had come to be used for the primordial matter out of which God created the universe, as well as for a confused mass, and shortly afterwards, a state of general confusion. *Paradise Lost* put these meanings together and

───

[6] John Milton, *Paradise Lost*, in *The Works of John Milton*, ed. Frank Allen Patterson *et al.*, 18 vols (New York: Columbia University Press, 1931–8), II.I (1931), Bk. VII, l. 168.

[7] Ibid., II. 931.   [8] Ibid., II. 829.   [9] Ibid., II. 948, 972, 974, 978.

even in the early nineteenth century, poets were still struggling to make imaginative use of the combination of fullness (confused, primordial matter) and emptiness (void, chasm, or abyss) in writing about Chaos. Or rather, perhaps, many of them were drawn to write about Chaos because it gave them the opportunity to try to make sense of the apparently contradictory but actually rather complementary properties of these two conditions of space.

The landscape in *Paradise Lost*'s Chaos is indeterminate, changing from sea to land; and even the underlying physical realities are undifferentiated and unpredictable. The abyss is 'without dimension, where length, breadth & heighth | And time and place are lost'.[10] Lawrence Babb comments that 'this passage cannot be taken literally [...] to mean that location, duration, and direction do not exist in Chaos. This is simply the expression of the superlative in confusion'.[11] But Milton goes to lengths to make sure that time and place are indeed lost within Chaos, not only for those characters who enter it but—so far as possible—for the reader as s/he travels through the Chaos passages. For example, he dramatizes the loss of time in the void by insisting on the loss of time in reading about it. When Satan is poised at the edge of the void contemplating his journey 'through the void immense', Milton begins 'Into this wild abyss,' then digresses with description for six lines, and has to repeat 'Into this wild abyss' again in order to be able to continue with the narrative:

> Into this wild abyss,
> The womb of Nature and perhaps her grave,
> Of neither sea, nor shore, nor air, nor fire,
> But all these in their pregnant causes mixed
> Confus'dly, and which thus must ever fight,
> Unless th'Almighty Maker them ordain
> His dark materials to create more worlds,
> Into this wild abyss the wary Fiend
> Stood on the brink of Hell and looked a while,
> Pondering his voyage;[12]

The description of the void has been so vivid and so mimetic that it has caused the reader to forget the direction of the narrative, and this is the point. Elizabeth Deeds Ermarth rightly comments that

[10] John Milton, *Paradise Lost*, in *The Works of John Milton*, II. 893.
[11] Lawrence Babb, *The Moral Cosmos of Paradise Lost* ([East Lansing]: Michigan State University Press, 1970), p. 107.
[12] Milton, *Paradise Lost*, II. 829, 910–19.

the whole poem produces the effect of simultaneity by shifting backwards and forwards in time, confirming the final apocalyptic view of historical sequence. [...] Even the language, with its inflected rhythms, suspends the ordinary sentence sequence and unifies the impression by paratactic means. These discontinuities in the medium distort the patterns and rhythms of ordinary perception [...].[13]

As Wordsworth described it, the model Alpine landscape at Lucerne allowed the spectator the visionary experience of comprehending and understanding a whole tract of country at once.[14] Similarly, *Paradise Lost*, in Ermarth's account, attempts to simulate the experience of perceiving all the moments of time in a single moment. The irregular temporality of the reader's experience of this poem is particularly important in those sections dealing with Chaos, where Milton subverts the linearity of reading time as part of his attempt to show how void space is a subversion of nature, a hiccup in the normal mapping of the world.

Paradise Lost gives us a sense of the cosmos radically different from that of the uniform, homogeneous field of nature underpinning mainstream early nineteenth-century thought, in which God's omnipresence guaranteed that no pockets of unreliability could exist, no part of space be inaccessible to the universal laws which human reason was beginning to perceive. Milton's cosmos has holes in its fabric, and this makes for drama and contrast in the narrative and in the texture of the poem. Early nineteenth-century inheritors of Milton's legacy in epic had to contend with this fundamental change in their own and their readers' understanding of the field of nature. For some, Milton gave legitimacy to their doubts about the uniformity of nature; others who insisted on the reliability of divinely created space had to contend with problems of imagery and narrative method.

Previous chapters have emphasized the role of space in giving (and policing) access to information and then in organizing knowledge once it is acquired. The remainder of this chapter, though, reverses this focus and reads some epic poems—now almost entirely forgotten—striving to reinvent disorder and disorganization within a culture which prided itself on moving towards the eradication of those conditions.

---

[13] Elizabeth Deeds Ermath, *Realism and Consensus in the English Novel* (Princeton, N.J.: Princeton University Press, 1983), p. 62.

[14] See above, p. 68.

CHAOS AND EARLY NINETEENTH-CENTURY
EPICS

Early in Disraeli's 1847 novel *Tancred*, the hero finds himself discussing a fashionable new book entitled *The Revelations of Chaos*—Disraeli's parody of the then anonymous *Vestiges of the Natural History of Creation*.[15] Tancred judges from the title that 'the subject is rather obscure', but Lady Constance enthuses: 'it is one of those books one must read. It explains everything [...] it is impossible to contradict anything in it. You understand, it is all science'.[16] The scene marks Tancred's disillusionment with Lady Constance, whom he had been thinking of marrying. Her disgraceful willingness to accept the tenets of evolutionary speculation and, almost worse, her stupid inability to reproduce them accurately ('Ah! that's it: we were fishes, and I believe we shall be crows') put her beyond the pale.[17] Disraeli's choice of title for this absurd but momentous scientific work illustrates the surprisingly high profile of 'Chaos' in early nineteenth-century literary debate about science, particularly the kind of all-embracing science, ranging from cosmology to biology, that Lady Constance's favourite book represents.

Chaos, a state of radically disordered matter and the utter negation of conceptual categorization, was a most useful site of experimentation for writers, particularly since, unlike Heaven and Hell, it had very limited theological significance by the early nineteenth century. Writers of the period who needed a place in which to situate arguments and anxieties about the various kinds of order and categorization that were being derived from—and imposed on—the physical and intellectual world often turned to 'Chaos'. On the whole in this period, 'chaos' is freighted in this period with unpleasant connotations, often associated with fear of dissolution. Bulwer Lytton's 1828 novel *Pelham*, for example, uses 'chaos' to indicate a state of psychological excitement so great that the compartments of rational thought give way: 'stormy thoughts, feelings, and passions so long at rest, rushed again into a terrible and

---

[15] James A. Secord briefly discusses this passage from *Tancred* as part of the reception of *Vestiges* in *Victorian Sensation: The Extraordinary Publication, Reception, and Secret Authorship of Vestiges of the Natural History of Creation* (Chicago: University of Chicago Press, 2000), pp. 188–90.

[16] Benjamin Disraeli, *Tancred: Or, the New Crusade* (London: Henry Colburn, 1847; repr. Longmans, 1871), pp. 109–10 (bk II, ch. 9).

[17] Ibid., p. 109 (bk II, ch. 9).

tumultuous action. The newly formed stratum of my mind was swept away; everything seemed a wreck, a chaos, a convulsion of jarring elements.'[18] Described from within, as here, chaos is experienced as pain and shock; from without, it is often represented as an inaccessible blank that resists the imagination. When Elinor and Harleigh in Fanny Burney's *The Wanderer* discuss the immortality of the soul, Elinor expresses astonishment that Harleigh can believe in the doctrine, since to her the subject is 'a chaos,—dark,—impervious, impenetrable!'.[19] Here 'chaos' suggests something too sublime to be comprehended, but elsewhere it was used in a much more hostile way. Charles Kingsley used 'chaos' to stand for something that cannot be understood because it is based on categories not sanctioned by God: 'Do but let the Bible tell its own story; grant, for the sake of argument, the truth of the dogmas which it asserts throughout, and it becomes a consistent whole. When a man begins, as Strauss does, by assuming the falsity of its conclusions, no wonder if he finds its premises a fragmentary chaos of contradictions.'[20] For Kingsley's Dean Winnstay, chaos is the opposite of the divinely underwritten totality and consistency of correct thought.

Frequently, chaos represents destruction, the threat of ultimate and irremediable disorder, an act of un-creation. In Maturin's 1820 bestseller *Melmoth the Wanderer*, a paean to the beneficent communication of humanity with nature contrasts with the apocalyptic silence of nature in its extreme places: 'the desolation now presented to the eyes of Immalee, was that which is calculated to cause terror, not reflection. Earth and heaven, the sea and the dry land, seemed mingling together, and about to replunge into chaos.'[21] In this instance chaos is the state of uncreated matter which we are more used to associating with the beginning than with the end of the world. By reasserting the existence of chaos at the end of nature, Maturin's metaphor imagines the history of the created universe as a mere blip in the much larger history of chaos, which is waiting to swallow it.

---

[18] Edward Bulwer Lytton, *Pelham; Or, The Adventures of a Gentleman* (1828) (London: Routledge, 1877), p. 365 (ch. 75).

[19] Fanny Burney, *The Wanderer; Or, Female Difficulties* (1814), ed. Margaret Anne Doody, Robert L. Mack and Peter Sabor (London: Penguin, 1991), p. 597 (book VII, ch. 63).

[20] Charles Kingsley, *Alton Locke, Tailor and Poet: An Autobiography* (1850), (London: 1881), p. 415 (ch. 38).

[21] Charles Maturin, *Melmoth the Wanderer: A Tale*, (1820), ed. Chris Baldick (Oxford: Oxford World's Classics, 1989), p. 321 (vol. III, ch. XVII).

But chaos could be used much more optimistically, as a place in which to regroup before launching on a new beginning, as when Disraeli's Coningsby, having spent a day 'in a dark trance rather than a reverie [...] he was like a particle of Chaos', re-emerges into London with a fresh sense of identity and self-worth and dedicates himself to public service.[22] Perhaps the best-known example of early nineteenth-century writing using Chaos/chaos as a creative rather than destructive space is *Frankenstein*. Shelley's Introduction to the 1831 edition explains that

invention, it must be humbly admitted, does not consist in creating out of void, but out of chaos; the materials must, in the first place, be afforded: it can give form to dark, shapeless substances, but cannot bring into being the substance itself. [...] Invention consists in the capacity of seizing on the capabilities of a subject, and in the power of moulding and fashioning ideas suggested to it.[23]

Chaos here is a storehouse of matter awaiting organization by the shaping power of an ordering and categorizing intelligence, very much as it is in the most favourable descriptions of Chaos in *Paradise Lost*. The parallel with Frankenstein's ordering intelligence which giving form to the 'dark, shapeless' (because largely undescribed) raw materials from the charnelhouse is obvious.

The two decades following the publication of *Frankenstein* saw a great surge in publication and sales of epic poems which, more directly than that novel, retell Biblical stories, especially the story of the creation. Although these poems not infrequently outsold the canonical early nineteenth-century poets, they have long since slipped from critical memory, and even in their own time were sometimes the subject of much derision. Contemporary reviewers used the popularity of this new genre as an opportunity to compare the standing of poetry in their own day with the masterworks of the past. The Biblical verse epic served usefully in critical accounts of the difference between high and middling poetry, and of the harmful effects the middling kind could have on the appreciation of the high. For example, Macaulay's review of Robert Montgomery's very popular epics (including *On the Omnipresence of the Deity* (1828) and *Satan* (1830)) lamented the fact that Montgomery's readership 'has been greater than that of Southey's Roderic, and beyond

[22] Benjamin Disraeli, *Coningsby; Or, The New Generation* (1844), ed. Thom Braun (Harmondsworth: Penguin, 1983), (book IX, ch. 4).

[23] Mary Shelley, 'Introduction', *Frankenstein: Or, The Modern Prometheus* (1831 edn), ed. M. K. Joseph (Oxford: Oxford World's Classics, 1969), p. 8.

all comparison greater than that of Cary's Dante, or of the best works of Coleridge'.[24] Similarly, the SDUK's publisher Charles Knight noted sadly in his memoirs that this 'was a period in which mediocrity was essentially necessary to great literary success. [*On the Omnipresence of the Deity*] went rapidly through five or six editions. The 'Excursion' had reached a second edition in ten years.'[25] Another damning review of Montgomery appeared in an early number of *Fraser's*, where he was lambasted not only for his own poems but also for encouraging the production of other worthless Biblical epics: 'We have the fatal proof [of Montgomery's influence] before us, in the shape of sundry poems, of ample thickness and balaam-weight, to wit, *Creation, Mount Sinai, The Impious Feast, Cain*, &c. &c. &c.'[26] Colin Graham argues that the 'act of literary salvage' that reading these and other such poems involves 'is always going to be in danger of reifying into an argument of popularity versus quality (the number of epics written against the fact that they are "bad" poetry'.[27] In their own day, these two factors were often counted as weighing on the same side. The large sales figures confirmed the low quality of the poetry. The tolerance of even Herbert Tucker is strained by the 'flabbergasting badness' of these poems, although 'no lover of Milton and Blake can reject such epic topics out of hand'.[28]

Despite the critical obloquy heaped on them in their own day and their almost total oblivion since, the poems do give an insight into the contemporary spatial imagination inflected through religious and cosmological speculation. For obvious reasons, those poems that tell the

[24] [T. B. Macaulay], review of Robert Montgomery's *The Omnipresence of the Deity: A Poem*, and *Satan, a Poem, Edinburgh Review*, 51 (April 1830), 193–210, p. 209. The references to Cary and Coleridge are linked by the fact that it was Coleridge who had made popular Cary's seminal but initially ignored translation of Dante. *On the Omnipresence of the Deity* seems to have gone into at least half a dozen editions in its first year; it was adapted for an oratorio in 1830 and a new edition produced for schools in the 1840s. *Satan* went into multiple editions within its first decade.

[25] Charles Knight, *Passages of a Working Life During Half a Century [...]*, 3 vols (London: Bradbury & Evans, 1864), II, p. 54.

[26] [William Maginn (?)], review of [John A. Heraud], *The Descent into Hell, Fraser's Magazine for Town and Country*, 1 (April 1830), 341–53 (p. 342). The works Maginn referred to were: William Ball, *Creation: A Poem* (London: Bull, 1830); William Phillips, *Mount Sinai: A Poem in Four Books* (London: Maunder, 1830); John E. Reade, *Cain The Wanderer, A Vision of Heaven, Darkness, and Other Poems* (London: Whittaker and Treacher, 1829) and Robert Eyres Landor, *The Impious Feast: A Poem in Ten Books* (London: Hatchard, 1828).

[27] Colin Graham, *Ideologies of Epic: Nation, Empire and Victorian Epic Poetry* (Manchester: Manchester University Press, 1998), p. 6.

[28] Herbert Tucker, 'Epic', in *A Companion to Victorian Poetry*, ed. Richard Cronin, Alison Chapman and Antony Harrison (Blackwell, 2002), pp. 25–41 (p. 35).

story of the creation of the world were generally the most imaginative and ambitious in their account of space and spaces, but most kinds of Biblical epic gave their writers the scope to describe the entire geography and history of the universe.

Contemporary with the surge in composition of Biblical epics, and seeming to stimulate yet further production of them, were John Martin's paintings and prints. Martin's hugely successful images of Biblical scenes, including his 1825–7 series of illustrations to *Paradise Lost*, formed a kind of visual equivalent to the more minor literary phenomenon of Biblical epic poems. Robert Montgomery's poem *Satan* describes the enormous impact of John Martin's *Paradise Lost* pictures. The Prince of Darkness, finding himself in early nineteenth-century London, comes upon one of Martin's scenes of Hell, and exclaims:

> The painter hath infernal pomp revealed.
> That second Milton, whose creative soul
> Doth shadow visions to such awful life,
> That men behold them with suspended breath,
> And grow etherial at a gaze!—how high
> And earthless hath his daring spirit soar'd,
> To paint the hell that kindled up the skies,
> And wield the lightning that his Maker hurl'd!

Montgomery's note to this passage asks 'To whom can this eulogium apply, but to the sublime painter of "Belshazzar's Feast," "The Deluge," "Fall of Nineveh," &c. &c.?' and exclaims 'Long may he live to adorn his age and country!'[29] The 'second Milton' is John Martin, identified by his paintings of Old Testament subjects, and praised by this polite and appreciative Satan whom Macaulay ridiculed as 'a respectable and pious gentleman'.[30] Martin's turbulent compositions, like many of his contemporaries' Biblical poems, used dramatic spatial imagery to represent the cosmology of the Christian universe, and to respond to the architectonics of the epic form. His canvases were typically dominated by a swirling vacant area near the centre of the composition. In 'The Creation of Light' from the *Paradise Lost* illustrations, for instance, Martin uses a thick, empty shaft of sunlight to create a cave-like effect in the midst of the surrounding landscape and clouds. At the heart of his oil sketch 'Satan Arousing the Fallen Angels', similarly, is a vast cavern

---

[29] R[obert] Montgomery, *The Poetical Works*, 5th edn, 3 vols (Glasgow: John Symington, 1839), I, pp. 128, 170.
[30] [T. B. Macaulay], review of *The Omnipresence of the Deity*, p. 209.

made by tongues of fire. Martin makes intense and sublime contrasts between occupied and unoccupied space, often pushing his figures and narratives out to the edges of the canvas while the centre is given over to various kinds of void space.

Martin's technical confidence put emptiness at the centre of his cosmological pictures; few of the poets producing Biblical epics at the same period had the equivalent of his mastery of composition. But again and again these poets took on the problem of making sense of emptiness. All the creation poems have to deal to some extent with the void space existing before the making of the world, and indeed some are careful to preserve pockets of that emptiness in the post-creation universe as well. The problems of describing void space and narrating events happening within it were inherited from Milton, but they could not be solved by imitating Miltonic methods, since few of these early nineteenth-century poets allowed themselves the luxury of entirely ignoring changes in the scientific understanding of the cosmos that had occurred over the intervening two centuries. Writing about the void generated literary, scientific, and theological problems for many of these poets, but it is clear that in some cases it was these problems that attracted the poets to Biblical epic in the first place. The darkness that Genesis describes as being on the face of the deep before the creation of light, the formless void itself, and Chaos and Night, later additions to this theological cosmology, are the negative spaces of the Judaeo-Christian universe. These pockets of disorder were fertile ground for writers wanting to unfold unorthodox theories that combined philosophy, science, and religion in varying proportions.

The task the writers of Biblical verse epic set themselves—that of narrating the past, present and future history of the borders of the spaces of the universe, including Heaven, Earth and Hell—obliged them to write about chaos as Chaos, not just a condition but also a specific location. In sophisticated hands this obligation drew forth elaborate and thoughtful imaginative constructions. In almost all cases, Chaos offered a space for philosophical, scientific and theological problems to be worked out in a way that could not be achieved within the ordered, legitimate, and consistent spaces of the divinely created universe.

The range of technical skill and sophistication of ideas evinced in the slew of Biblical verse epics of the 1820s and 30s was very wide. Some writers could see that the difficulties of exploring these negatives spaces were too great for them, and bypassed the problem completely. A. Gomershall's 1824 poem *Creation*, very possibly the most banal

example of the Biblical epic genre, dismissed the perplexities of the
universe prior to creation in two jaunty couplets:

> Envelop'd in chaotic darkness laid
> The vast expanse, till light God's voice obey'd.
> Let there be light the great Jehovah said,
> And instantly chaotic darkness fled![31]

The preface to this work explains that the author's motivation for
publishing was purely financial. Following bereavements and illnesses,
the author, now in her seventies, relied pathetically on subscriptions
to the poem to relieve her poverty. But where the author's reason for
publishing was to promote a quasi-philosophical thesis, the anomalous
spaces of the cosmos generally received more attention.

The anonymous author of the 1834 poem *The Wonders of Chaos and
the Creation Exemplified* was one of those who took advantage of the
disordered spaces of his imagined cosmos to set out a pet theory. He
explained in a prose preface that the purpose of his poem was to argue
that Chaos was caused by the fall of Lucifer. Because of the unorthodoxy
of this view he was at pains to emphasize his religious conformity: 'I am
desirous that every one who reads this Poem may perfectly understand
that my tenets are those of the established church.'[32] Chaos, for this
poet, represented an unacceptable interruption in the divinely ordered
universe. God would not have allowed 'such a confused heap as Chaos'
to have existed until it was necessary. If Chaos had been part of God's
initial design it would have suggested that the creation was a work of
'fancy or caprice' instead of 'triumph and victory'.[33] And, he argued, it
was clear that Lucifer's fall must have caused 'some great convulsion'
in the cosmic order. Accordingly, the poem explains in the blank verse
favoured by most of these poets, that at the time of Lucifer's and his
supporters' fall,

> tumult fill'd immensity around,
> Where all had been a vacuum before;
> For Chaos held no occupation there
> Until the fate of this rebellious band
> Of angels, but was a peaceful blank,
> By sin untenanted and undefiled.[34]

---

[31] A. Gomershall, *Creation, A Poem* (Newport: printed for the author, 1824), p. 4.
[32] [Anon.], *The Wonders of Chaos and the Creation Exemplified, A Poem, in Eight
Cantos* (London: Hatchard, 1834), pp. 5, 9.
[33] Ibid., pp. 5, 6.   [34] Ibid., p. 19.

But tears shed by the falling angels produce a 'putrifying and chaotic slime' in which Chaos forms. The organic metaphor of 'slime' is part of a strongly corporeal discourse which this poem uses to describe Chaos. At one point Chaos figures as a kind of perturbed digestive system:

> bubbling billows form, which bursting belch
> Dense and offensive steam; while roaring winds
> Sweep o'er their waves with unrestricted howl
> And aid confusion.[35]

and elsewhere as a disgusting and threatening feminine body, which God eventually redeems by making it the material for the creation of the earth:

> noisy Chaos, which to him [Lucifer] owers birth,
> No longer shall an useless mass remain
> Of crude defilement, an unsightly spot
> To my pure vision; and offensive doubt
> Of my controlment o'er immensity.
> The black and fulsome deep, which in its womb
> With sick'ning throes conceives corroding mire
> My pow'r shall cause from out its pregnant breast
> To yield an universe of glowing spheres,
> Lit by mine eye, which through profundity
> Shall blaze with light unquenchable, and add
> New territory to my boundless rule.[36]

So appalling is the disgust provoked by these corporeal images of chaos that only the creation of a world and the human race can blot it out. For this poet, the creation of the earth was not undertaken in order to produce humanity but in order to put an end to the affront of Chaotic space.

An anonymously published 1822 work called *Creation: A Poem* was one of the most interesting in terms of its approach to genre. Where the poet of *The Wonders of Chaos* had added a preface to offer his Church of England credentials, the author of this Creation poem exhibited a much freer attitude to genre, turning to Biblical epic to support a thesis he also advanced in the same year in a prose treatise titled

---

[35] Ibid., p. 25.
[36] Ibid., pp. 20, 26–7. The image of Chaos as 'pregnant' with the materials of the universe comes, of course, from *Paradise Lost*, II. 913.

Primum Mobile; Or, Solar Repulsion.[37] This treatise argued that the light of the sun is the origin of all motion in the solar system, as well as being—and here the author alluded to the 'fluid' hypothesis about the nature of electricity—the cause of electrical and magnetic phenomena.[38] His bizarre argument that 'a fluid proceeding from the sun, which is supposed to be of the nature of primitive fire, is the origin of the phenomena of electricity, galvanism, and magnetism' was, he admitted, 'not capable of absolute demonstration'.[39] Failing absolute demonstration by scientific argument, he turned to Biblical epic. The attempt to promote his argument using this form caused him some literary difficulties, as he himself acknowledged: 'the poetry will be found deficient chiefly in the more philosophical parts of the work, where the difficulty of blending science with poetry, rendered the execution naturally defective'.[40] Under these circumstances, his adoption of Biblical epic was a reflection of the genre's perceived cultural authority, as well as of its capacious ability to accommodate an extraordinary variety of ideological and even disciplinary stances.

Like the others in the genre, this poem builds on Genesis's account of Chaos as pre-existing God's creation of the universe.

> Darkness, and night, and chaos, reigned at first
> Throughout the deep, and spread their gloomy veil
> Impenetrable [...]
> And with their solitude and sleep of death,
> Sat dismal o'er the rude and unformed mass
> Of waters, that extended far and wide,
> And occupied the space on every side,
> Sole tenant of the vast ethereal void.[41]

These waters are the main focus of this poet's 'philosophical' interest, but no sooner are they named than he is obliged to enter a caveat about the limits of language in discussing them:

> Waters, the name this mountain-mass obtained
> [...]
> The nature of the fluid mass unknown,

---

[37] *Primum Mobile; or, Solar Repulsion; Being a Query Concerning the Primary Cause of Motion in the Solar System, as Connected with Gravity. By the Author of 'Creation', a Poem* (Liverpool: printed for the author, 1822).

[38] Ibid., p. 118.   [39] Ibid., p. 118.

[40] *Creation: A Poem. By the Author of 'Primum Mobile'* (Liverpool: printed for the author, 1822), p. i.

[41] Ibid., pp. 7–8.

> No true similitude existing now;
> Though still the name of water is retained
> For oceans or for rivers, pure or salt,
> Unlike the watery waste embodied then;[42]

The problem here is partly a version of the Miltonic difficulty of using post-lapsarian language to describe a pre-lapsarian world, except that in this poem it is not the Fall but the Creation itself that marks the barrier to translation. But a further linguistic concern is the familiar nineteenth-century problem of writing science in a language that includes terms in general, non-scientific, use. This writer adopts Genesis's term—'waters'—but means by the word something very different from 'water', and the opposing tugs from Biblical and scientific linguistic authority cannot be negotiated except via this rather clumsy explanation.

Ultimately his aim is to show that

> All motion, if we err not, owes to light
> Its primitive and first-born impulse, else
> Would matter rest inert, nor will to move.
> Light first disturbs the gravitating force,
> And turns the sleep of nature from its bed.[43]

Light is the product of the 'waters' that covered space before the creation, which may have been composed of some combustible material:

> Perhaps this sea of waters might contain
> Some liquid substances inflammable,
> Analogous to naphtha, such as flames
> In Baku's field, beside the Caspian sea,
> Or oleaginous petroleum,
> Which Eastern Ava's wells for lamp-light yield.[44]

When God ignites them to produce light, these petroly waters boil and give off steam, out of which God creates the universe. The steam is the subject of further linguistic and scientific anxiety. God calls it Firmament, but to humans it is known as ether, and its nature 'mortals cannot hope to know', except that it permits the transmission of 'radiant heat and light'.[45] As we saw in the previous chapter, the early nineteenth century was a period of major scientific controversy over the nature of ether, its interactions with gravity, and its role in propagating light.

---

[42] Ibid., p. 8.   [43] Ibid., p. 16.   [44] Ibid., p. 13.   [45] Ibid., p. 20.

The author of this poem was attempting to intervene in an enormously complex series of arguments in contemporary physics, arguments in which his prose treatise *Primum Mobile* had not given him a footing. The poem is a very interesting example of the use of Biblical poetry as a debating chamber in which writers outside the limits of the scientific establishment (as it was developing in the early decades of the century) attempted to substitute the cultural authority of epic for the disciplinary authority of science. The poet clearly recognizes that the demands of the two kinds of authority are in conflict with one another, and this is one reason why his language is under so much strain and his poetry becomes 'naturally defective'. Herbert Tucker is quite right to argue that early nineteenth-century religious epic manifested a 'drive to classify and dispose', but the classifications it offered did not always centre on the theological 'division of the wicked from the blessed'.[46] Here we have an instance of a poet whose classifications are struggling to bring together the special linguistic practices of cutting-edge science with the hallowed tradition of Biblical terminology in a task of edifying hopelessness which nonetheless indicates the possibilities for non-monologism in epic poetry.

In contrast with this author's ambitious scheme of mixing religious with scientific speculation, William Ball's 1830 *Creation* eschewed attempts to explain the nature of the cosmos in even quasi-scientific terms. Instead, the poem resorted to a mythological system which disagreed with most of its contemporaries by portraying the divinely ordained universe not as the norm but as an anomalous kind of space. Before the creation, most of space is the dominion of a figure sometimes referred to as 'Night' and sometimes 'Nought':

> There stands th'unshaken throne of primal Nought,
> And through all distance and all height, all depth,
> His fearful realms extend; save, here, a space,
> Scooped out from his dominions, wherein dwells
> A power amazing and unknown[47]

This power is God, who shares space amicably with his neighbour. Night is passive in the face of God's incursion on his territory: he

> sees, with tranquil eye, a universe
> Invade his endless realms: his secret throne,

---

46 Tucker, 'Epic', p. 35.
47 William Ball, *Creation: A Poem* (London: Bull, 1830), pp. 48–9.

> In motionless security, abides
> Within the gulf of darkness uttermost;
> Invincibly quiescent.[48]

Though it is made perfectly clear that God is greater than Night/Nought, this poem is unusual in seeing space as largely outside the boundaries of the divinely created universe.[49] The space that Night/Nought rules over is still and passive, an 'untroubled empire of repose', but though inert it does seem to be far bigger than God's realm.[50]

Ball's poem is very unusual in reversing the spatial organization of the cosmos in this way. Much more common were poems which saw Chaos as a local exception to the general rule of God's order. Indeed, some insisted that there were no exceptions at all to that rule. Contemporary Biblical epics sometimes agreed with the prose tracts attacking materialist cosmologies; these tracts generally asserted, as one of them put it, that 'were there a boundary to [God's] presence, there would be a limit to his ubiquity', and that such a limit was impossible to an all-powerful God.[51] Poets in non-epic genres were also apt to contribute to this debate about the limits of God's power in space; elevated meditations on nature like David Moir's 'Starlight Reflections' used chaos not as part of a realized mythological cosmology but as the antithesis of divinely ordered nature:

> How awful is the might of *Him*
>   Who stretch'd the skies from pole to pole!
> And breathed, through chaos waste and dim,
>   Creation's living soul!
> A thousand worlds are glowing round,
>   And thousands more than sight can trace
> Revolve throughout the vast profound,
>   And fill the realms of space:[52]

For Moir chaos is strictly in the past. Other poets, though, imagined chaos as perhaps still in existence, somewhere unimaginably distant from the created world. Thomas Moore's poem 'The Loves of the Angels' describes the delight taken by the Second Angel in watching the stars. As we saw in Chapter 3, astronomy was often in this period represented

---

[48] Ibid., p. 50.   [49] Ibid., p. 49.   [50] Ibid., p. 48.

[51] John Dudley, *The Anti-Materialist; Denying the Reality of Matter, and Vindicating the Universality of Spirit* (London: Bell, 1849), p. 19.

[52] David Macbeth Moir, 'Starlight Reflections', in *The Poetical Works of David Macbeth Moir*, 2 vols (Edinburgh: Blackwood, 1852), I, 105–10 (p. 105).

as the most morally elevating of the sciences, and Moore's poem is no exception:

> Innocent joy! alas, how much
> Of misery had I shunn'd below,
> Could I have still liv'd blest with such;
> Nor, proud and restless, burn'd to know
> The knowledge that brings guilt and woe.[53]

Studying the stars, the youthful angel flies in imagination

> in quest
> Of those, the farthest, loneliest,
> That watch, like winking sentinels,
> The void, beyond which Chaos dwells;[54]

Moore distinguishes 'the void' from Chaos but places both beyond the scope of our universe. But the theological problems raised by the possible contemporaneous existence of Chaotic and divinely-created space were too much for more orthodox poets who stressed that God's rule extended through all space. Robert Montgomery's massively successful poem *On the Omnipresence of the Deity* was typical of this wing of the argument, addressing God as

> Pervading SPIRIT, whom no eye can trace,
> Felt through all time, and working in all space,
> Imagination cannot paint that spot,
> Around, above, beneath, where Thou art not.[55]

(The faint echo of Donne's 'Elegy: To his Mistress Going to Bed' — 'Licence my roving hands, and let them go | Behind, before, above, between, below' — is surely accidental: Montgomery's is not a sensual religious poetry.[56])

The Biblical epic that worked hardest to think through the artistic problems of writing about truly empty space was John Abraham Heraud's *The Descent into Hell*. Published anonymously in 1830, near the start of the genre's brief heyday, *The Descent into Hell* tells the story of Christ's activities during the three days between the crucifixion and the

---

[53] Thomas Moore, 'Second Angel's Story' in *The Loves of the Angels* (1823), *The Poetical Works*, 10 vols (London: Longman *et al.*, 1840–1), IX, p. 40.

[54] Moore, 'Second Angel's Story', p. 41.

[55] Robert Montgomery, *On the Omnipresence of the Deity: A Poem*, 4th edn (London: Maunder, 1828), p. 17.

[56] John Donne, 'Elegy 19: To His Mistress Going to Bed', in *The Complete English Poems*, ed. A. J. Smith (Harmondsworth: Penguin, 1971), pp. 124–6 (p. 125).

resurrection. Heraud explicitly sets out to distinguish his cosmological descriptions from Milton's, and nowhere is his vision more clearly differentiated from that of *Paradise Lost* than in his account of Chaos. For Heraud, Chaos is a state of mind as much as a place. Like Hell, which he argued is 'the place of separate spirits in general not [...] any place of torment in particular', Chaos is subjective:

> Each tells the truth, yet each in telling lies,
> Of that far region, which no region is,
> All mystery, yet hath no mysteries—
> Far region, ever yet at hand, I wis,
> Within us and about us every-where—[57]

Michael Wheeler points out that though controversial, the view that both heaven and hell should be understood as states of mind rather than real places was widely held in the Victorian period.[58] Chaos, of course, had a far less canonical status in the Christian cosmos than either heaven and hell and though, as we have seen, it could still present theological difficulties for early nineteenth-century Anglicans, it was more open to innovative imaginative treatments than the central locations of Protestant theology. Accordingly, Heraud was free to create a Chaos that really was a void. Unlike Milton's, which contained undifferentiated matter, Heraud's Chaos containing nothing at all. Indeed, Heraud seems to be rejecting Miltonic ideas of chaotic space when he announces that

> They feign who tell of rocks abrupt and dread,
> Of precipice, and waste outrageous deep
> Of waters, in an agonizing bed,
> That sweat with torture while they madly sweep
> With sounds of human voice! Still vacancy,
> Void o'er whose formless face doth darkness sleep,
> Is all the way, beyond the boundary
> Of temporal space[59]

Heraud, re-imagining Milton's Christian cosmology for the nineteenth century, refuses to allow void space to disrupt his organized universe. Instead of Milton's accounts of perilous turbulence and indiscriminate

---

[57] [John A. Heraud], *The Descent into Hell: A Poem* (London: Murray, 1830), p. 79.
[58] Michael Wheeler, *Heaven, Hell, and the Victorians* (Cambridge: CUP, 1994), pp 188–9.
[59] [Heraud], *The Descent Into Hell*, p. 78.

matter, Heraud offers a 'pure space' of 'empty forms and shapeless shades', which makes for

> A quiet voyage, whence whoso dissuades,
> With tales of tumult or detours of pain,
> With Fancy peoples and with Sense pervades.[60]

In this passage Heraud once again uses the challenges that void space poses to narrative and to cosmology as an opportunity for differentiating his own vision from Milton's. But the operations of time are much more simple in Heraud's poem than in Milton's. He makes an attempt at a Miltonic sense of simultaneity in his description of lost spirits who wander in the void

> Millions of years no measure can define,
> Past in an instant, as the lightning is—
> And was—[...][61]

But neither in sentence construction nor in chronological shifts does Heraud's poem match Ermarth's description of the experience of time in reading *Paradise Lost*. Milton's poem arranges the elements of its narrative and its cosmology into a nonchronological array which is representative at once of humanly perceived disorder and divinely ordained order. Heraud's poem merely asks the reader to be a witness to the events narrated.

The absolute stillness and vacancy of Heraud's void means that no time can pass there. Like many philosophical writers, Heraud considers time to be the succession of events, and in the void there are no events, not even the quotidian celestial events of the sun's and moon's progress:

> No Sun is here to measure o'er the sky
> Day; Moon nor Stars, to rule the night, or tell
> Of seasons; here is no variety
> Of Time, nor Time himself.[62]

This nontemporal world presents great difficulties for the poet working with necessarily temporal verbal language. So Heraud invents a narrator who can create a kind of temporality out of 'the well | Of [his] own being':

> a pure sphere of light
> I can project, and shape and syllable

---

[60] Ibid., pp. 78, 79.   [61] Ibid., p. 80.
[62] Ibid., p. 81.

> With Form and Name; or on the darkness drear,
> E'en as the eye of Childhood doth, create
> Picture and friezes indistinct or clear.
> These may poetic fancy aggregate
> In her own time and space, eath as the sense
> Of Euclid could construct and demonstrate
> Ideas, as his own intelligence
> Perfect and pure, by power of his own mind,
> Shaped by its prescript, and proceeding thence.[63]

The imagination can use such projections to produce an artificial time and space through which the reader will be enabled to experience the world of the void. In a move that would be quite characteristic of this period's non-fiction prose but is more unusual in verse, this process of production of temporality and spatiality is likened to an idealized version of Euclid's development of the principles of geometry.

Herschel's and Brougham's accounts of mathematics used the image of a man working out mathematical truths from first principles without resort to sensory experimentation.[64] Like them, Heraud considers an ability to think geometrically to be an intrinsic property of the mind, albeit one exemplified by the remarkably gifted mind of Euclid. Heraud uses the term 'sense' to describe this ability or gift, as if the mental processes involved in geometrical imaginings were a kind of physical sense additional to the bodily ones. This use of 'sense' reflects Heraud's difficulties with narrating nontemporal and nonspatial 'events' in a language which can only deal in the temporal and spatial. Heraud tries to solve the problem of rendering events in terms of sense while at the same time emphasizing the metaphysical and paradoxical aspects of a journey which defies human senses:

> The dying Saint, with a calm smile,
> So, the same instant, leaves this world beneath,
> And reaches th'other, passing no defile,
> Of toil or travel; with his farewell breath,
> Smoothly transported to a blessèd goal:
> Of Past or Future no account with Death.
> All indivisible as his own Soul,
> Eternity broods o'er the Infinite,
> Time has no lapse and Space is one and whole.

---

[63] Ibid., p. 81.   [64] See above, pp. 94–5.

> Therefore it was, his transit on my sight
> Glanced and was gone, returning through the void
> To his far home[65]

Heraud insists that neither time nor space intervened in the journey, and yet is forced to concede that time and space were sufficiently involved for the traveller to be visible, however briefly, to the watching narrator. His effort to destabilize the categories of space and time is impaired by his obligation to narrate events as if they took place within those categories. By making his void truly empty, and not, like Milton's, empty only of organization, Heraud poses himself the insoluble problem of imagining events happening within a nontemporal and nonspatial realm. The extreme dematerialization of the void in Heraud's poem gives no purchase for narrative. His was the most ambitious and innovative depiction of Chaos among the nineteenth-century Biblical epicists and the route he mapped out was not followed by others.

So one of the central concerns for the writers of cosmological epic was whether Chaos represented the norm in space and time and the divinely created universe the anomaly, or vice versa. The relationship between the two was almost always seen as an opposition, notwithstanding the often suggested possibility that the one was necessary for the existence of the other. The two kinds of space represented the struggle between order and disorder, meaning and non-meaning. Early nineteenth-century middle-brow poets seem to have brooded over the problem of the legitimate role of Chaos in the universe and the results to have sold in very large quantities, despite the critical obloquy heaped on them in the periodicals. It is tempting to see the genre itself, as well as the Chaotic spaces it enumerates, as (in Foucault's term) heterotopic—as an 'other space' in which the laws applying to the systems beyond it have no power. Heterotopias, the cultural geographer Kevin Hetherington argues,

bring together heterogeneous collections of unusual things without allowing them a unity or order established through resemblance. Instead, their ordering is derived from a process of similitude which produces, in an almost magical, uncertain space, monstrous combinations that unsettle the flow of discourse [...]. A further principle of heterotopia, one that derives from their significance

---

[65] Heraud, *The Descent into Hell*, p. 82.

as representing through similitude, is that they only exist in relation, that is, they are established by their difference in a relationship between sites rather than their Otherness deriving from a site itself. It is not the relationship within a space that is the source of this heterotopic relationship, for such an arrangement, seen from within that space, may make perfect sense. It is how such a relationship is seen from outside, from the standpoint of another perspective, that allows a space to be seen as heterotopic.[66]

Milton's void does seem to fit this characterization of heterotopia, in that from within, Chaos has indeed a certain internal logic and consequent navigability; it is from without that it seems to pose such a challenge to the ordering of the cosmos. John Heraud, on the other hand, re-imagined the void as silent, still, and nontemporal as well as nonspatial. This kind of void both resists and encourages identification as heterotopic, since it certainly appears different from the ordering of the spaces which surround it, and yet it does not seem to order items within itself; a visiting narrator is required to provide even the spatio-temporal framework which precedes linguistic ordering.

Hetherington describes heterotopias as 'almost like laboratories' in that they 'can be taken as the sites in which new ways of experimenting with ordering society are tried out'.[67] The suggestion of a laboratory-like space seems to invite investigation of the void spaces of physical science as heterotopia. The possibility of empty space was a contested and controversial area of physical science from Aristotle through to the early nineteenth century. In the science of Milton's time, the discovery of potentially void 'Torricellian' space renewed this old controversy, so that vying in *Paradise Lost* are a number of theological, scientific, and metaphysical arguments about the role of emptiness in a divinely created universe. By the early nineteenth century, vacua had come to be accepted as forming part of the fabric of the world; but the development of 'classical' field theory by Faraday, Thomson, and Maxwell, entailed a revision of the category of 'empty' space so that in the physical sciences of the 1850s and after it was no longer simple to distinguish bodies from the space in which they existed.

Chaos and voids are spaces which cannot be read, because they are empty of matter or of distinction; yet they offer reassuring proof of

---

[66] Kevin Hetherington, *The Badlands of Modernity: Heterotopia and Social Ordering* (London: Routledge, 1997), p. 43.

[67] Ibid., p. 13.

the existence of a system which can be read. All verbally expressed systems necessarily impose a kind of spatiality and temporality on the material they organize and on the authority they claim. By exploring the blank spaces within those systems, early nineteenth-century readers and writers could perceive anew the connections and the disjunctions that made knowledge and authority possible.

Afterword

This book describes a very particular moment in the history of British scientific culture. The interactions of literature and science in the 1820s, 30s and 40s were made excitable, urgent, and energetic by the sense that control was being lost: that the pace of change, in both the content and the nature of the disciplines, was becoming too rapid for the existing structures of dissemination and assimilation of knowledge to cope with. By some time around 1860, even the attempt to cope had been given up. On the whole literature and science settled into considering one another as definitively different systems of knowledge, with no individual expected to master both.

To follow the themes of this book into British literary and scientific culture beyond the mid-1850s, then, would require a different method of reading and argument. For one thing, the kind of evidence I have concentrated on here—writing mainly by people who thought of science and literature as essentially mutually friendly because sharing common aims and a common response to contemporary conditions—is much less characteristic of the second half of the nineteenth century than of the first. For another, the disciplinary boundaries not only of science but also of literature were so much more clearly delineated in mid- and late-Victorian Britain that a substantial gap begins to open between the discussions occupying science and the filtering of those discussions through to the general and literary reader. Of course the publication of *Origin of Species* in 1859 was immediately followed by widespread and sustained popular, literary, and wider cultural debate. But in the hard sciences, the shift to routine use of a language dominated not by words but by mathematics, the increasing remoteness of scientific procedures from popular consciousness, and the solidifying of structures for producing and authenticating knowledge meant that even so crucial a development as field theory took decades to emerge into literary culture.

Of course these two factors—the changing relationship of literature and science, and the growing inaccessibility of science's languages—are intimately linked. Political and social change also profoundly affected the interactions of science in culture: as attitudes to mass education shifted in ways we can see reflected in the Forster Education Act of 1870, state provision for basic scientific education for the working classes replaced a previous generation's improvized and informal adventures in autodidacticism. Both on the national political stage, and in the institutional politics of the disciplines themselves, later nineteenth-century British science was far more organized and official than its counterpart of fifty years earlier. And its place in British culture was enormously altered by its increasingly distinct and discrete identity.

Generalizations of the kind are satisfying but profoundly unsteady, liable to destabilization by myriads of counter-examples and exceptions. To reconsider this trajectory of scientific cultural history from the point of view of Victorian women, for example, affects some of its assumptions. And yet, while there is an immediately recognizable difference not just in the content but also in the tone, the mode of address and the rhetorics of authority between a scientific publication of the 1810s and another in the same discipline from the 1840s, the difference between a paper in physics from Faraday's mid-career and another from Kelvin's is even greater. The way in which science located itself in broader culture can be measured by such differences.

The three decades following the end of the Long French Wars were an extraordinary moment in the cultural history of British science. To read that moment through the lens of writing about space allows us to investigate the territorialities, the negotiations of contextualization, the endless disturbances and resettlings in contemporary conceptions of knowledge, which characterize the period's response to mass education and a scientific revolution. But we still do not know enough about Victorian understandings of space. Though pathbreaking work has been done to uncover various kinds of (mainly urban) spatial practice, there is a great deal more to be done before we can understand the full range of the period's spatial imagination. This book is a beginning; but we have yet fully to explore the cultural life of the imaginary, the hypothetical, and the abstract spaces in which no nineteenth-century person walked, but with and through which they thought.

Bibliography

PRIMARY SOURCES

Ampère, André-Marie, *Essai sur la philosophie des sciences, ou exposition analytique d'une classification naturelle de toutes les connaissances humaines* (Paris: Bachelier, 1834).

Anon., 'Kyan's Patent—The Nine Muses,—and the Dry-Rot', *Bentley's Miscellany*, 1 (1837), 93–6.

Anon., *Creation: A Poem. By the Author of 'Primum Mobile'* (Liverpool: printed for the author, 1822).

Anon., *Primum Mobile; or, Solar Repulsion; Being a Query Concerning the Primary Cause of Motion in the Solar System, as Connected with Gravity. By the Author of 'Creation', a Poem* (Liverpool: printed for the author, 1822).

Anon., *The Wonders of Chaos and the Creation Exemplified, A Poem, in Eight Cantos* (London: Hatchard, 1834).

Arnott, Neil, *Elements of Physics, or Natural Philosophy General and Medical, Explained Independently of Technical Mathematics, and Containing New Disquisitions and Practical Suggestions*, 3rd edn (London: Longman *et al.*, 1828).

Babbage, Charles, *Reflections on the Decline of Science in England* (London: Fellowes and Booth, 1830; repr. Farnborough: Gregg, 1969).

_____ *The Ninth Bridgewater Treatise: A Fragment* (London: John Murray, 1837).

Ball, W. W. Rouse, *A History of the Study of Mathematics at Cambridge* (Cambridge: CUP, 1889).

Ball, William, *Creation: A Poem* (London: Bull, 1830).

Bamford, Samuel, *Passages in the Life of a Radical* (1844) (London: MacGibbon & Kee, 1967).

Bentham, Jeremy, *Chrestomathia: Being a Collection of Papers, Explanatory of the Design of an Institution, Proposed to be Set on Foot, Under the Name of The Chrestomathic Day School [...] for the Use of the Middling and Higher Ranks of Life* (London: Payne and Foss, 1816).

_____ *The Works of Jeremy Bentham*, ed. John Bowring, 11 vols (Edinburgh: Tait, 1843).

Berzelius, Jacob, *Jahres-Bericht über die Fortschritte der physischen Wissenschaften von Jacob Berzelius: Eingereicht an die schwedische Akademie der Wissenschaften, den 31 März 1832*, trans. F. Wöhler (Tubingen: Laupp, 1833).

Bickersteth, Edward Henry, *Yesterday, To-Day, and For Ever: A Poem, in Twelve Books* (London: Rivingtons, 1866).

Bompass, Charles Carpenter, *An Essay on the Nature of Heat, Light, and Electricity* (London: Underwood, 1817).

[Brougham, Henry], review of Thomas Young, *Bakerian Lecture on the Theory of Light and Colours, Edinburgh Review*, 1:2 (Jan 1803), 450–6.

——*A Discourse of the Objects, Advantages, and Pleasures of Science*, 2nd edn (London: Baldwin, Cradock, and Joy, 1827).

——'Practical Observations upon the Education of the People', in *Speeches of Henry Lord Brougham* ... , 4 vols (Edinburgh: Black *et al.*, 1838), IV, 99–152.

Bulwer Lytton, Edward, *Pelham; Or, The Adventures of a Gentleman* (1828) (London: Routledge, 1877).

Burke, Edmund, *A Philosophical Inquiry into Our Ideas of the Sublime and Beautiful* (1757), ed. David Womersley (London: Penguin, 1998).

Burney, Fanny, *The Wanderer; Or, Female Difficulties* (1814), ed. Margaret Anne Doody, Robert L. Mack and Peter Sabor (London: Penguin, 1991).

Chalmers, Thomas, *The Efficacy of Prayer Consistent with the Uniformity of Nature* (London: Partridge and Oakey, 1849).

Chambers, E., *Cyclopaedia; or An Universal Dictionary of Arts and Sciences; [...]*, 2nd edn, corrected and amended, with additions, 2 vols (London: Midwinter and others, 1738).

Chambers, W. and R., *Introduction to the Sciences: For Use in Schools and for Private Instruction* (Edinburgh: Chambers, 1838).

——*A Description and Guide to the British Metropolis; Embossed by Permission for the Use of the Blind* (Glasgow: John Smith *et al.*, 1841).

——*Introduction to the Science of Astronomy, Embossed by Permission, for the Use of the Blind* (Glasgow: John Smith *et al.*, 1841).

Chevallier, Temple, *The Study of Mathematics as Conducive to the Development of the Intellectual Powers* (Durham: Parker, 1836).

Coleridge, Samuel Taylor, 'Preliminary Treatise on Method', *Encyclopedia Metropolitana*, 5th edn (1851).

——*Aids to Reflection* (1824), ed. John Beer (London: Routledge and Princeton, N.J., Princeton University Press, 1993).

——*Philosophical Lectures of Samuel Taylor Coleridge*, ed. Kathleen Coburn (London: Pilot Press, 1949).

——*Collected Letters of Samuel Taylor Coleridge*, ed. Earl Leslie Griggs, 6 vols (Oxford: Clarendon Press, 1956–71).

——*The Notebooks of Samuel Taylor Coleridge*, ed. Kathleen Coburn, 2 double vols (London: Routledge & Kegan Paul, 1957–62).

——*The Friend*, ed. Barbara Rooke, *Collected Works of Samuel Taylor Coleridge*, 4, 2 vols (London: Routledge & Kegan Paul, 1969).

——*Biographia Literaria; or, Biographical Sketches of My Literary Life and Opinions*, ed. James Engell and W. Jackson Bate, 2 vols (London: Routledge & Kegan Paul, 1983).

Cottle, Joseph, *Malvern Hills, with Minor Poems*, 4th edn (London: Cadell, 1829).

Craig, Edward, *A Lecture on the Formation of a Habit of Scientific Enquiry, Delivered at the Staines Literary Institution* (Staines: Smith, 1836).

Daubeny, Charles, *An Introduction to the Atomic Theory, Comprising a Sketch of the Opinions Entertained by the Most Distinguished Ancient and Modern Philosophers With Respect to the Constitution of Matter* (Oxford and London: Murray, 1831).

——*An Introduction to the Atomic Theory*, 2nd edn, (Oxford: University Press, 1850).

[De Morgan, Augustus], 'English Science', *British and Foreign Review*, 1.1 (1835), 134–57.

De Quincey, Thomas, *The Works of Thomas De Quincey*, gen. ed. Grevel Lindop, 21 vols (London: Pickering and Chatto, 2000–3).

Dick, Thomas, *On the Improvement of Society by the Diffusion of Knowledge: or, An Illustration of the Advantages which would Result from a More General Dissemination of Rational and Scientific Information Among All Ranks* (Edinburgh: Waugh and Innes; Dublin: Curry; London: Whittaker, 1833).

Dickens, Charles, *The Life and Adventures of Nicholas Nickleby* (1839), ed. Mark Ford (London: Penguin, 1999).

——*The Life and Adventures of Martin Chuzzlewit* (1844), ed. Patricia Ingham (London: Penguin, 1999).

——*Our Mutual Friend* (1865), ed. Adrian Poole (London: Penguin, 1997).

Disraeli, Benjamin, *Coningsby; Or, The New Generation* (1844), ed. Thom Braun (Harmondsworth: Penguin, 1983).

——*Tancred: Or, the New Crusade* (London: Henry Colburn, 1847: repr. Longmans, 1871).

Donne, John, *The Complete English Poems*, ed. A. J. Smith (Harmondsworth: Penguin, 1971).

Drummond, William Hamilton, *The Pleasures of Benevolence: A Poem* (London: Wakeman, 1835).

Dudley, John, *The Anti-Materialist; Denying the Reality of Matter, and Vindicating the Universality of Spirit* (London: Bell, 1849).

Eastlake, Charles L., *Hints on Household Taste: The Classic Handbook of Victorian Interior Design*, 4th rev. edn, (London: Longman, Green, 1878; repr. Mineola, N.Y: Dover, 1986).

Eddington, Arthur, *The Philosophy of Physical Science* (Cambridge: CUP, 1939).

Eliot, George, *The Mill on the Floss* (1860), ed. Gordon S. Haight (Oxford: Clarendon Press, 1980).

——*Middlemarch* (1871–2) ed. David Carroll (Oxford: Clarendon Press, 1986).

[Faraday, Michael], 'On Imagination and Judgement', in 'Mental Exercises', Royal Institution, Faraday MSS, F. 13.A, pp. 22–7.

[Faraday, Michael], 'On the Pleasures and Uses of the Imagination', in 'Mental Exercises', Royal Institution, Faraday MSS, F. 13.A, pp. 39–47.

Faraday, Michael, *Michael Faraday in Wales: Including Faraday's Journal of his Tour through Wales in 1819*, ed. Dafydd Tomos ([Denbigh (?)]: Gwasg Gee, [1972]).

———— *Experimental Researches in Chemistry and Physics* (London: Taylor and Francis, 1859).

———— *The Correspondence of Michael Faraday*, ed. Frank A. J. L. James (London: Institution of Electrical Engineers, 1991–).

Finleyson, John, *The Universe As It Is, and the Detection and Refutation of Sir Isaac Newton; Also, the Exposure and Proved Fabrication of the Solar System* (London: Snell, 1830).

Flavel, John, *Husbandry Spiritualized; or, The Heavenly Use of Earthly Things […]* (London: Boulter, 1669).

[Fonblanque, Albany], review of Robert Montgomery's *Satan*, *Westminster Review*, 12 (April 1830), 355–8.

Freneau, Philip Morin, *Poems*, ed. Harry Hayden Clark (New York: Harcourt, Brace, 1929).

Gaskell, W. P., *An Address to the Operative Classes, being the Substance of a Lecture Explanatory and in Defence of the Nature and Objects of the Cheltenham Mechanics' Institution […]* (Cheltenham: Gray, 1835).

[Grove, William], 'Physical Science in England', *Blackwood's Edinburgh Magazine*, 54 no. 336 (October 1843), 514–25.

Gomershall, A., *Creation, A Poem* (Newport: printed for the author, 1824).

Hazlitt, William, *The Selected Writings of William Hazlitt*, ed. Duncan Wu, 9 vols (London: Pickering & Chatto, 1998).

Hemans, Felicia, *The Works of Mrs. Hemans*, 7 vols (Edinburgh: Blackwood, 1839).

[Heraud, John A], *The Descent into Hell: A Poem* (London: Murray, 1830).

Heraud, John A, *The Judgement of the Flood* (London: Fraser, 1834).

———— *The In-Gathering […]* (London: Simpkin, Marshall, 1870).

Herschel, John F. W., *A Preliminary Discourse on the Study of Natural Philosophy* (London: Longman, 1830).

———— *Letter from J. F. W. Herschel, Esq.* (No place of publication, no publisher: 1830[?]).

———— 'Light', *Encyclopaedia Metropolitana*, 29 vols (London: Fellowes, 1830), IV, pp. 341–586.

———— 'Presidential Address', *British Association for the Advancement of Science Reports* (1846), xxviii–xliv.

[Herschel, J. F. W.], review of Humboldt's *Kosmos*, *Edinburgh Review*, 87 (January 1848), 170–229.

Higgins, W. M., *An Introductory Treatise on the Nature and Properties of Light, and on Optical Instruments* (London: Nimmo; Edinburgh: Blackwood; Dublin: Vurry, 1829).

Hood, Thomas, *The Complete Poetical Works of Thomas Hood*, ed. Walter Jerrold (London: Frowde and Oxford University Press, 1906).

Humboldt, Alexander von, *Cosmos: A Sketch of a Physical Description of the Universe*, trans. E. C. Otté *et al.*, 5 vols (London: Bohn, 1849–65).

_____ *Cosmos: Sketch of a Physical Description of the Universe*, trans. under the superintendence of Edward Sabine, new edn, 5 vols (London: Longman and John Murray, 1856).

Jewsbury, Geraldine, *The Half Sisters* (1848) (Oxford: Oxford World's Classics, 1994).

Kant, Immanuel, *Theoretical Philosophy 1755–1770*, trans. and ed. David Walford in collaboration with Ralf Meerbote (Cambridge: CUP, 1992).

Keats, John, *The Letters of John Keats 1814–1821*, ed. by Hyder Edward Rollins, 4 vols (Cambridge: CUP, 1958).

Kingsley, Charles, *Alton Locke, Tailor and Poet: An Autobiography* (1850) (London: Macmillan, 1881).

Knight, Charles, *Passages of a Working Life During Half a Century […]*, 3 vols (London: Bradbury & Evans, 1864).

Kyan, John Howard, *On the Elements of Light, and their Identity with those of Matter, Radiant and Fixed* (London: Longman *et al.*, 1838).

Lardner, Dionysius, *The First Six Books of The Elements of Euclid […]* (London: Taylor, 1828).

_____ *A Treatise on Geometry, and Its Application in the Arts* (London: Longman *et al.*, 1840).

Lyell, Charles, *Principles of Geology*, ed. James A. Secord (Harmondsworth: Penguin, 1997).

[Macaulay, T. B.], review of Robert Montgomery's *The Omnipresence of the Deity: A Poem*, and *Satan, a Poem*, *Edinburgh Review*, 51 (April 1830), 193–210.

MacConnell, Thomas, *A Lecture on the Signs of the Times: Delivered in the Great Lecture Room of Robert Owen's Institution, Gray's Inn Road* (London: Eamonson *et al.*, 1832).

Mackintosh, T. S., *The 'Electrical Theory' of the Universe; Or, The Elements of Physical and Moral Philosophy* (London: Simpkin, Marshall; and Manchester, Heywood [c. 1838]).

Macvicar, John G., *Inquiries Concerning the Medium of Light and the Form of its Molecules* (Edinburgh: Black; London: Longman *et al.*, 1833).

[Maginn, William (?)], review of [John A. Heraud], *The Descent into Hell*, *Fraser's Magazine for Town and Country*, 1 (April 1830), 341–53.

[Maginn, William (?)], review of Heraud's *The Judgement of the Flood, Fraser's Magazine for Town and Country*, 9 (May 1834), 522–34.

Malthus, Thomas, *Principles of Political Economy* (1820), ed. John Pullen, 2 vols (Cambridge: CUP, 1989).

Mann, John, *A Glance at the Objects of Thought, or, A Concise and Systematic View of the Different Branches of Human Knowledge* (London: Roake and Varty, 1833).

Marx, Karl and Friedrich Engels, *The Communist Manifesto* (1848), trans. Samuel Moore (1888), ed. Gareth Stedman Jones (London: Penguin, 2002).

Maturin, Charles, *Melmoth the Wanderer: A Tale*, (1820), ed. Chris Baldick (Oxford: Oxford World's Classics, 1989).

Maxwell, James Clerk, *The Scientific Papers of James Clerk Maxwell*, ed. W. D. Niven, 2 vols (Cambridge: CUP, 1890; repr. Mineola, N.Y.: Dover, 2003).

Milton, John, *The Works of John Milton*, ed. Frank Allen Patterson *et al.*, 18 vols (New York: Columbia University Press, 1931–8).

Moir, David Macbeth, *The Poetical Works of David Macbeth Moir*, 2 vols (Edinburgh: Blackwood, 1852).

[Moll, Gerard], *On the Alleged Decline of Science in England, By a Foreigner* (London: Boosey, 1831).

Montgomery, Robert, *On the Omnipresence of the Deity: A Poem*, 4th edn (London: Maunder, 1828).

—— *The Poetical Works*, 5th edn, 3 vols (Glasgow: John Symington, 1839).

Moore, Thomas, *The Poetical Works*, 10 vols (London: Longman *et al.*, 1840–1).

More, Hannah, *The Works of Hannah More*, 6 vols (London: Fisher, Fisher and Jackson, 1834).

Newman, Francis William, *Theism, Doctrinal and Practical or, Didactic Religious Utterances* (London, 1858).

Newton, Isaac, *Mathematical Principles of Natural Philosophy and his System of the World*, trans. Andrew Motte, rev. Florian Cajori, 2 vols (Berkeley: University of California Press, 1966).

Nicol, James, *Introductory Book of the Sciences, Adapted for the Use of Schools and Private Students* (Edinburgh: Oliver and Boyd; and London: Simpkin, Marshall, 1844).

Novalis, *The Novices of Sais*, trans. Ralph Manheim (New York: Valentin, 1949).

Oersted, Hans Christian, *The Soul in Nature: With Supplementary Contributions*, trans. Leonora and Joanna B. Horner (London: Bohn, 1852).

Olson, Charles, *Collected Prose*, ed. Donald Allen and Benjamin Friedlander (Berkeley: University of California Press, 1997).

Payne Knight, Richard, *An Analytical Inquiry into the Principles of Taste*, 2nd edn (London: Payne and White, 1805).

Peacock, Thomas Love, *The Works of Thomas Love Peacock [...]*, ed. Henry Cole, 3 vols (London: Bentley, 1875).

Phillips, William, *Mount Sinai: A Poem in Four Books* (London: Maunder, 1830).

Planché, J. B., *The Bee and the Orange Tree; Or, The Four Wishes: An Original Fairy Extravaganza in One Act* (London: no publ., 1846).

Plato, *Meno*, in *The Dialogues of Plato [...]*, trans. Benjamin Jowett, 4 vols (Oxford: Clarendon Press, 1871), I, pp. 247–92.

—— *Timaeus*, trans. Benjamin Jowett, in *The Dialogues of Plato [...]*, 4 vols (Oxford: Clarendon Press, 1871), II, pp. 513–86.

Powell, Baden, *The Present State and Future Prospects of Mathematical and Physical Studies in the University of Oxford [...]* (London: Parker, 1832).

—— *A General and Elementary View of the Undulatory Theory [...]* (London: Parker, 1841).

Prescot, B., *The Inverted Scheme of Copernicus; with the Pretended Experiments upon which his Followers have Founded their Hypotheses of Matter and Motion, Compared with Facts, and with the Experience of the Senses: and the Doctrine of the Formation of Worlds out of Atoms, by the Power of Gravity, Contrasted with the Formation of One World by Divine Power, as it is Revealed in the History of the Creation* (Liverpool and London: Riebau, 1822).

'Pry, Paul', *The Blunders of a Bigwig [...]* (London: Hearne, 1827).

Reade, John E., *Cain The Wanderer, A Vision of Heaven, Darkness, and Other Poems* (London: Whittaker and Treacher, 1829).

Ritchie, William, *Principles of Geometry, Familiarly Illustrated, and Applied to a Variety of Useful Purposes, Designed for the Instruction of Young Persons* (London: Taylor, 1833).

Shelley, Mary, *Frankenstein: Or, The Modern Prometheus* (1831 edn), ed. M. K. Joseph (Oxford: Oxford World's Classics, 1969).

Smiles, Samuel, *Self-Help; with Illustrations of Character, Conduct, and Perseverance* ed. Peter W. Sinnema (Oxford: Oxford World's Classics, 2002).

Smith, Alfred, *An Introductory Lecture on the Past and Present State of Science, In this Country, As Regards the Working Classes* (Ripon: Langdale; London: Longman *et al.*, 1831).

Somerville, Mary, *On the Connexion of the Physical Sciences* (London: John Murray, 1834).

—— *Physical Geography*, 2 vols (London: John Murray, 1848).

—— *Personal Recollections, from Early Life to Old Age, of Mary Somerville*, ed. Martha Somerville (London: John Murray, 1873).

Sprat, Thomas, *The History of the Royal-Society of London, for the Improving of Natural Knowledge* (London: Martyn & Allestry, 1667).

Stuart-Wortley, Emmeline, *Ernest Mountjoy, A Comedietta* (London: n. publ., 1844).

Tennyson, Alfred, *The Poems of Tennyson*, ed. Christopher Ricks, 3 vols (Harlow: Longman, 1969; 2nd edn 1987).

Thackeray, William Makepeace, *Vanity Fair* (1848), ed. John Carey (London: Penguin, 2001).

Wheatley, George, *An Address on the Necessity, Uses, and Advantage of Affording to the Labouring Classes, the Means of Acquiring General, Scientific, Moral, and Political Knowledge [...]* (Whitehaven: Cook, 1832).

Whewell, William, *An Essay on Mineralogical Classification and Nomenclature; with Tables of the Orders and Species of Minerals* (Cambridge: Smith, 1828).

_____ 'On the Employment of Notation in Chemistry', *Journal of the Royal Institution of Great Britain*, 1 (1830–1), 437–53.

_____ *Address Delivered in the Senate-House at Cambridge, June XXV, MDCCCXXXIII. On the occasion of the opening of the third General Meeting of the British Association for the Advancement of Science* (Cambridge: Smith, 1833).

[Whewell, William], review of Mary Somerville's *On the Connexion of the Physical Sciences*, in *Quarterly Review*, 51 (March 1834), 54–68.

Whewell, William, *Astronomy and General Physics, Considered with Reference to Natural Theology*, 4th edn (London: Pickering, 1834).

_____ *Thoughts on the Study of Mathematics, as Part of a Liberal Education* (Cambridge: Deighton; London: Whittaker, 1835).

_____ *History of the Inductive Sciences, from the Earliest to the Present Times*, 3 vols (London: Parker; and Cambridge: Deighton, 1837).

_____ *The Mechanical Euclid, Containing the Elements of Mechanics and Hydrostatics Demonstrated after the Manner of the Elements of Geometry [...]* (Cambridge: Deighton; London: Parker, 1837).

_____ *The Philosophy of the Inductive Sciences, Founded Upon Their History*, 2 vols (London: Parker, 1840).

_____ 'On the Influence of the History of Science upon Intellectual Education', in *Lectures on Education: Delivered at the Royal Institution of Great Britain* (London: Parker, 1855), pp. 3–36.

Wordsworth, William, 'Preface to the Lyrical Ballads' (1802), in *Lyrical Ballads*, ed. R. L. Brett and A. R. Jones (London: Routledge, 1963, 2nd edn 1991), pp. 241–72.

_____ *A Guide through the District of the Lakes in the North of England, with a Description of the Scenery, &c. for the use of Tourists and Residents*, 5th edn (Kendal: Hudson and Nicholson; London: Longman and others, 1835), in *The Prose Works of William Wordsworth*, ed. W. J. B. Owen and Jane Worthington Smyser, 3 vols (Oxford: Clarendon Press, 1974), II, pp. 121–465.

_____*Shorter Poems, 1807–1820*, ed. by Carl H. Ketcham (Ithaca, N.Y.: Cornell University Press, 1989).

_____*The Prelude*, (1805), in *The Prelude: 1799, 1805, 1850*, ed. Jonathan Wordsworth, M. H. Abrams and Stephen Gill (New York: Norton, 1979).

SECONDARY SOURCES

Andrews, Malcolm, *The Search for the Picturesque: Landscape Aesthetics and Tourism in Britain, 1760–1800* (Aldershot: Scolar, 1989).

Armstrong, Tim, 'Poetry and Science', in Neil Roberts, ed., *A Companion to Twentieth-Century Poetry* (Oxford: Blackwell, 2001), pp. 76–88.

Astore, William J., *Observing God: Thomas Dick, Evangelicalism, and Popular Science in Victorian Britain and America* (Aldershot: Ashgate, 2001).

Babb, Lawrence, *The Moral Cosmos of Paradise Lost* ([East Lansing]: Michigan State University Press, 1970).

Barfield, Owen, *What Coleridge Thought* (London: Oxford University Press, 1971).

Barrell, John, *The Dark Side of the Landscape: The Rural Poor in English Painting 1730–1840* (Cambridge: CUP, 1980).

Bate, Jonathan, *Romantic Ecology: Wordsworth and the Environmental Tradition* (London: Routledge, 1991).

Beer, Gillian, *Darwin's Plots: Evolutionary Narrative in Darwin, George Eliot and Nineteenth-Century Fiction* (London: Routledge and Kegan Paul, 1983; 2nd edn, Cambridge: CUP, 2000).

_____*Open Fields: Science in Cultural Encounter* (Oxford: Clarendon Press, 1996).

Bermingham, Ann, *Landscape and Ideology: The English Rustic Tradition, 1740–1860* (London: Thames and Hudson, 1986).

Blakemore, M. J., and J. B. Harley, 'Concepts in the History of Cartography: A Review and Perspective', *Cartographica* monographs 26, *Cartographica* 17.4, (Winter 1980), 1–120.

Brattin, Joel J., 'Constancy, Change, and the Dust Mounds of *Our Mutual Friend*', *Dickens Quarterly*, 19 (2002), 23–30.

Brock, W. H., 'The Selection of the Authors of the Bridgewater Treatises', *Notes and Records of the Royal Society of London*, 21:2 (1966), 162–79.

Brown, Daniel, *Hopkins' Idealism: Philosophy, Physics, Poetry* (Oxford: Clarendon Press, 1997).

Buchwald, Jed Z., *The Rise of the Wave Theory of Light: Optical Theory and Experiment in the Early Nineteenth Century* (Chicago: University of Chicago Press, 1989).

Burnett, John, David Vincent and David Mayall, eds, *The Autobiography of the Working Class: An Annotated, Critical Bibliography*, 3 vols (London: Harvester, 1984–9).

Buzard, James, *The Beaten Track: European Tourism, Literature, and the Ways to Culture, 1800–1918* (Oxford: Clarendon Press, 1993).

Cantor, Geoffrey *Optics after Newton: Theories of Light in Britain and Ireland, 1704–1840* (Manchester: Manchester University Press, 1983).

——*Michael Faraday: Sandemanian and Scientist* (Houndmills: Macmillan, 1991).

——'The Scientist as Hero: Public Images of Michael Faraday', in Michael Shortland and Richard Yeo, eds., *Telling Lives in Science: Essays on Scientific Biography* (Cambridge: CUP, 1996), pp. 171–94.

Cantor, G. N. and M. J. S. Hodge, eds, *Conceptions of Ether: Studies in the History of Ether Theories 1740–1900* (Cambridge: CUP, 1981).

Certeau, Michel de, *The Practice of Everyday Life*, trans. Steven Rendall (Berkeley: University of California Press, 1984).

Chard, Chloe, and Helen Langsdon, eds., *Transports: Travel, Pleasure, and Imaginative Geography, 1600–1830* (New Haven, Ct.: Yale University Press, 1996).

Clifford, James, *Fieldnotes: The Makings of Anthropology*, ed. Roger Sanjek (Ithaca, N.Y.: Cornell University Press, 1990).

Colie, Rosalie L., *Paradoxia Epidemica: The Renaissance Tradition of Paradox* (Princeton, N.Y.: Princeton University Press, 1966).

Connor, Stephen, 'Voice, Technology and the Victorian Ear', in *Transactions and Encounters: Science and Culture in the Nineteenth Century*, ed. Roger Luckhurst and Josephine McDonagh (Manchester: Manchester University Press, 2002), pp. 16–29.

Copley, Stephen, and Peter Garside, eds., *The Politics of the Picturesque: Literature, Landscape and Aesthetics since 1770* (Cambridge: CUP, 1994).

Cosgrove, Denis, 'Prospect, Perspective and the Evolution of the Landscape Idea', *Transactions of the Institute of British Geographers*, 10 (1985), 45–62.

Crawford, Rachel, *Poetry, Enclosure, and the Vernacular Landscape, 1700–1830* (Cambridge: CUP, 2002).

Crosland, Maurice P., *Historical Studies in the Language of Chemistry* (London: Heinemann, 1962).

Darrigol, Olivier, *Electrodynamics from Ampère to Einstein* (Oxford: Oxford University Press, 2000).

Derrida, Jacques, 'Khora', trans. Ian McLeod, in *On the Name*, ed. Thomas Dutoit (Stanford, Calif.: Stanford University Press, 1995).

Duhem, Pierre, *Medieval Cosmology: Theories of Infinity, Place, Time, Void and the Plurality of Worlds*, ed. and trans. Roger Ariew (Chicago: University of Chicago Press, 1985).

Ellis, Kate Ferguson, *The Contested Castle: Gothic Novels and the Subversion of Domestic Ideology* (Urbana: University of Illinois Press, 1989).

Ermarth, Elizabeth Deeds, *Realism and Consensus in the English Novel* (Princeton, N.J.: Princeton University Press, 1983).

Everett, Nigel, *The Tory View of Landscape* (New Haven, Ct.: Yale University Press, 1994).

Fisch, Menachem, and Simon Schaffer, eds., *William Whewell: A Composite Portrait* (Oxford: Clarendon Press, 1991).

Flanders, Judith, *The Victorian House: Domestic Life from Childbirth to Deathbed* (London: HarperCollins, 2003).

Flint, Kate, 'Dust and Victorian Vision', in Juliet John and Alice Jenkins, eds, *Rethinking Victorian Culture* (Basingstoke: Macmillan, 2000), pp. 46–62.

_____ *The Victorians and the Visual Imagination* (Cambridge: CUP, 2000).

Foucault, Michel, *Power/Knowledge: Selected Interviews and Other Writings, 1972–1977*, ed. by Colin Gordon, trans. Colin Gordon *et al.* (Brighton: Harvester, 1980).

_____ *The Foucault Reader*, ed. Paul Rabinow (Harmondsworth: Penguin, 1984).

_____ 'Of Other Spaces', *Diacritics*, 16 (1986), 22–7.

Friedman, Michael, *Kant and the Exact Sciences* (Cambridge, Ma.: Harvard University Press, 1992).

Fulford, Tim, *Landscape, Liberty and Authority: Poetry, Criticism and Politics from Thomson to Wordsworth* (Cambridge: CUP, 1996).

Genette, Gérard, *Figures of Literary Discourse*, trans. Alan Sheridan (Oxford: Blackwell, 1982).

Gifford, Don, *The Farther Shore: A Natural History of Perception, 1798–1984* (London: Faber and Faber, 1990).

Gillispie, Charles Coulston, *Genesis and Geology: A Study in the Relations of Scientific Thought, Natural Theology and Social Opinion in Great Britain, 1790–1850* (London: Oxford University Press, 1951).

Gooday, Graeme, 'Faraday Reinvented: Moral Imagery and Institutional Icons in Victorian Electrical Engineering', *History of Technology* 15 (1993), 190–205.

Graham, Colin, *Ideologies of Epic: Nation, Empire and Victorian Epic Poetry* (Manchester: Manchester University Press, 1998).

Greenfield, Bruce, 'The Problem of the Discoverer's Authority in Lewis and Clark's *History*', in *Macropolitics of Nineteenth-Century Literature: Nationalism, Exoticism, Imperialism*, ed. Jonathan Arac and Harriet Ritvo (Philadelphia: University of Pennsylvania Press, 1991), pp. 12–36.

Gregory, Derek, *Geographical Imaginations* (Oxford: Basil Blackwell, 1994).

Guralnick, Stanley M., 'The Contexts of Faraday's Electrochemical Laws', *Isis*, 70 (1979), 59–75.

Haggett, Peter, *Locational Analysis in Human Geography* (London: Arnold, 1965).

Hamilton, James, *A Life of Discovery: Michael Faraday, Giant of the Scientific Revolution* (London: Random House, 2002).

Handy, Ellen, 'Dust Piles and Damp Pavements: Excrement, Repression, and the Victorian City in Photography and Literature', in Carol T. Christ and John O. Jordan, eds, *Victorian Literature and the Victorian Visual Imagination* (Berkeley: University of California Press, 1995), pp. 111–33.

Haraway, Donna, *Simians, Cyborgs and Women: The Reinvention of Nature* (London, Free Association Books, 1991).

Harley, J. B., 'Maps, Knowledge, and Power', in Denis Cosgrove and Stephen Daniels, eds, *The Iconography of Landscape: Essays on the Symbolic Representation, Design and Use of Past Environments* (Cambridge: CUP, 1988), pp. 277–303.

Harman, P. M., *Energy, Force, and Matter: The Conceptual Development of Nineteenth-Century Physics* (Cambridge: CUP, 1982).

_____ *The Natural Philosophy of James Clerk Maxwell* (Cambridge: CUP, 1998).

Harvey, David, 'The Geopolitics of Capitalism', in *Social Relations and Spatial Structures*, ed. Derek Gregory and John Urry (Basingstoke: Macmillan, 1985), pp. 128–63.

Heidegger, Martin, 'The Age of the World Picture', in *The Question Concerning Technology, and other Essays*, trans. William Lovitt (New York: Harper and Row, 1977), pp. 115–54.

Hesse, Mary, *Forces and Fields: The Concept of Action at a Distance in the History of Physics* (London: Nelson, 1961).

Hetherington, Kevin, *The Badlands of Modernity: Heterotopia and Social Ordering* (London: Routledge, 1997).

Howson, Geoffrey, *A History of Mathematics Education in England* (Cambridge: CUP, 1982).

Jenkins, Alice, 'Spatial rhetoric in the self-presentation of nineteenth-century scientists: Faraday and Tyndall', in Crosbie Smith and Jon Agar, eds., *Making Space for the History of Science* (Basingstoke: Macmillan, 1998), pp. 181–91.

_____ 'Writing the Self and Writing Science: Mary Somerville as Autobiographer', in Juliet John and Alice Jenkins, eds, *Rethinking Victorian Culture* (Houndmills: Macmillan, 2000), pp. 163–79.

Johnson, Richard, 'Educational Policy and Social Control in Early Victorian England', *Past And Present* 49 (November 1970), 96–119.

_____ ' "Really Useful Knowledge": Radical Education and Working-Class Culture, 1790–1848', in John Clarke, Chas Critcher and Richard Johnson, eds, *Working-Class Culture: Studies in History and Theory* (London: Hutchinson, 1979), pp. 75–102.

Joyce, Simon, *Capital Offenses: Geographies of Class and Crime in Victorian London* (Charlottesville: University of Virginia Press, 2003).

King, Geoff, *Mapping Reality: An Exploration of Cultural Cartographies* (Basingstoke: Macmillan, 1996).

Knight, David, *The Age of Science: The Scientific World-view in the Nineteenth Century* (Oxford: Basil Blackwell, 1986).

_____ *Humphry Davy, Science and Power* (Cambridge: CUP, 1992).

Kristeva, Julia, *Revolution in Poetic Language*, trans. Margaret Waller (New York: Guildford, 1984).

Lefebvre, Henri, *The Production of Space*, trans. Donald Nicholson-Smith (Oxford: Basil Blackwell, 1991).

Levere, Trevor H., *Poetry Realized in Nature: Samuel Taylor Coleridge and Early Nineteenth-century Science* (Cambridge: CUP, 1981).

Lewis, Judith S., 'Princess of Parallelograms and Her Daughter: Math and Gender in the Early Nineteenth Century Aristocracy', *Women's Studies International Forum*, 18 (1995), 387–94.

Livingstone, David N.,'The Spaces of Knowledge: Contributions Towards a Historical Geography of Science', *Environment and Planning D: Society and Space*, 13 (1995), 5–34.

Logan, Thad, *The Victorian Parlour* (Cambridge: CUP, 2001).

Luckhurst, Roger, *The Invention of Telepathy* (Oxford: Oxford University Press, 2002).

Marcus, Sharon, *Apartment Stories: City and Home in Nineteenth-Century Paris and London* (Berkeley: University of California Press, 1999).

Massey, Doreen, *Space, Place and Gender* (Cambridge: Polity, 1994).

Mingay, G. E., *Enclosure and the Small Farmer in the Age of the Industrial Revolution* (London: Macmillan, 1968).

Mitchell, Timothy, *Colonising Egypt* (Cambridge: CUP, 1988).

Moretti, Franco, *Atlas of the European Novel, 1800–1900* (London: Verso, 1998).

Morrell, Jack, and Arnold Thackray, *Gentlemen of Science: Early Years of the British Association for the Advancement of Science* (Oxford: Clarendon Press, 1981).

Morus, Iwan Rhys, *Frankenstein's Children: Electricity, Exhibition, and Experiment in Early Nineteenth-Century London* (Princeton, N.J.: Princeton University Press, 1998).

——*Michael Faraday and the Electrical Century* (Cambridge: Icon, 2004).

Nead, Lynda, 'Mapping the Self: Gender, Space and Modernity in Mid-Victorian London', in Roy Porter, ed., *Rewriting the Self: Histories from the Renaissance to the Present* (London: Routledge, 1997), pp. 167–85.

——*Victorian Babylon: People, Streets, and Images in Nineteenth-Century London* (New Haven, Ct.: Yale University Press, 2000).

Neeson, J. M., *Commoners: Common Right, Enclosure and Social Change in England, 1700–1820* (Cambridge: CUP, 1993).

Nersessian, Nancy J., *Faraday to Einstein: Constructing Meaning in Scientific Theories* (Dordrecht: Martinus Nijhoff, 1984).

Newey, Katherine, 'Attic Windows and Street Scenes: Victorian Images of the City on the Stage', *Victorian Literature and Culture*, 25.2 (1997), 253–62.

Noakes, Richard, 'Spiritualism, Science and the Supernatural in mid-Victorian Britain', in Nicola Bown, Carolyn Burdett and Pamela Thurschwell, eds, *The Victorian Supernatural* (Cambridge: CUP, 2004), pp. 23–43.

Nord, Deborah Epstein, *Walking the Victorian Streets: Women, Representation, and the City* (Ithaca, N.Y.: Cornell University Press, 1995).

Nye, Mary Jo, *From Chemical Philosophy to Theoretical Chemistry: Dynamics of Matter and Dynamics of Disciplines, 1800–1950* (Berkeley: University of California Press, 1993).

Oppenheim, Janet, *The Other World: Spiritualism and Psychical Research in England, 1850–1914* (Cambridge: CUP, 1985).

Otis, Laura, 'Introduction', Laura Otis, ed., in *Literature and Science in the Nineteenth Century: An Anthology* (Oxford: Oxford University Press, 2002).

Patten, Robert L., 'From House to Square to Street: Narrative Traversals', in Helena Michie and Ronald R. Thomas, eds., *Nineteenth-Century Geographies: The Transformation of Space from the Victorian Age to the American Century* (New Brunswick, N.J.: Rutgers University Press, 2003), pp. 91–206.

Pettitt, Clare, *Patent Inventions: Intellectual Property and the Victorian Novel* (Oxford: Oxford University Press, 2004).

Pike, David L., 'Underground Theater: Subterranean Spaces on the London Stage', *Nineteenth Century Studies*, 13 (1999), 102–38.

Pratt, Mary Louise, *Imperial Eyes: Travel Writing and Transculturation* (London: Routledge, 1992).

Radford, Andrew, *Thomas Hardy and the Survivals of Time* (Aldershot: Ashgate, 2003).

Rappaport, Erika Diane, *Shopping for Pleasure: Women in the Making of London's West End* (Princeton, N.J.: Princeton University Press, 2000).

Rauch, Alan, *Useful Knowledge: The Victorians, Morality, and the March of Intellect* (Durham, N.C.: Duke University Press, 2001).

Restivo, Sal, *Mathematics in Society and History: Sociological Inquiries* (Dordrecht: Kluwer, 1992).

Richards, Joan L., *Mathematical Visions: The Pursuit of Geometry in Victorian England* (San Diego, Calif.: Academic Press, 1988).

Richardson, Alan, *Literature, Education, and Romanticism: Reading as Social Practice, 1780–1832* (Cambridge: CUP, 1994).

Robbins, Michael, *The Railway Age* (London: Routledge & Kegan Paul, 1962; 2nd edn, Manchester: Manchester University Press, 1998).

Sachs, Aaron, 'The Ultimate "Other": Post-Colonialism and Alexander von Humboldt's Ecological Relationship with Nature', *History and Theory*, 42 (2003), 111–35.

Said, Edward, *The World, the Text, and the Critic* (Cambridge, Ma.: Harvard University Press, 1983; repr. London: Faber and Faber, 1984).

Sallis, John, *Chorology: On Beginning in Plato's Timaeus* (Bloomington: Indiana University Press, 1999).

Secord, James A., *Controversy in Victorian Geology: The Cambrian-Silurian Dispute* (Princeton: Princeton University Press, 1986).

——— *Victorian Sensation: The Extraordinary Publication, Reception, and Secret Authorship of Vestiges of the Natural History of Creation* (Chicago: University of Chicago Press, 2000).

Shapin, Steven, and Simon Schaffer, *Leviathan and the Air-Pump: Hobbes, Boyle, and the Experimental Life* (Princeton, N.J.: Princeton University Press, 1985).

Shelden, Pamela J., 'Jamesian Gothicism: The Haunted Castle of the Mind', *Studies in the Literary Imagination*, 7 (1974), 121–34.

Silver, Harold, *The Concept of Popular Education: A Study of Ideas and Social Movements in the Early Nineteenth Century* (London: MacGibbon and Kee, 1965).

Smith, Crosbie, 'Force, Energy and Thermodynamics', in *The Cambridge History of Science*, vol. 5, *The Modern Physical and Mathematical Sciences*, ed. Mary Jo Nye (Cambridge: CUP, 2003), pp. 289–310.

Sneiders, H. A. M., 'Oersted's Discovery of Electromagnetism', in Andrew Cunningham and Nicholas Jardine, eds, *Romanticism and the Sciences* (Cambridge: CUP, 1990), pp. 228–40.

Soja, Edward W., *Postmodern Geographies: The Reassertion of Space in Critical Social Theory* (London: Verso, 1989).

Stafford, Barbara, *Voyage into Substance: Art, Science, Nature, and the Illustrated Travel Account, 1760–1840* (Cambridge, Ma.: MIT Press, 1984).

Stott, Anne, *Hannah More: The First Victorian* (Oxford: Oxford University Press, 2003).

Swenson, Loyd Sylvan, Jr, *The Ethereal Aether: A History of the Michelson-Morley-Miller Aether-Drift Experiments, 1880–1930* (Austin: University of Texas Press, 1972).

Talon, Henri, 'Space, Time, and Memory in *Great Expectations*', *Dickens Studies Annual*, 3 (1974), 122–33.

Topham, Jonathan R., 'Beyond the "Common Context": The Production and Reading of the Bridgewater Treatises', *Isis*, 89:2 (1998), 233–62.

Thompson, E. P., *The Making of the English Working Class* (London: Gollancz, 1963).

Tuan, Yi-Fu, 'Language and the Making of Place: A Narrative-Descriptive Approach', *Annals of the Association of American Geographers*, 81 (1991), 684–96.

Tucker, Herbert, 'Epic', in *A Companion to Victorian Poetry*, ed. Richard Cronin, Alison Chapman and Antony Harrison (Blackwell, 2002), pp. 25–41.

Varey, Simon, *Space and the Eighteenth-Century English Novel* (Cambridge: CUP, 1990).

Vincent, David, *Bread, Knowledge and Freedom: A Study of Nineteenth-Century Working-Class Autobiography* (London: Europa, 1981).

Walkowitz, Judith R., 'Going Public: Shopping, Street Harrassment, and Streetwalking in Late Victorian London', *Representations*, 62 (1988), 1–30.
____ *City of Dreadful Delight: Narratives of Sexual Danger in Late-Victorian London* (London: Virago, 1992).

Ward, Graham, *The Postmodern God: A Theological Reader* (Oxford: Blackwell, 1997).

Watkins, Eric, ed., *Kant and the Sciences* (Oxford: Oxford University Press, 2001).

Webb, R. K., *The British Working Class Reader, 1790–1848: Literacy and Social Tension* (London: Allen & Unwin, 1955, repr. New York: Kelley, 1971).

Wheeler, Michael, *Heaven, Hell, and the Victorians* (Cambridge: CUP, 1994).

Wiley, Michael, *Romantic Geography: Wordsworth and Anglo-European Spaces* (Houndmills: Macmillan, 1998).

Williams, L. Pearce, *Michael Faraday: A Biography* (London: Chapman and Hall, 1965).

Williams, Raymond, *The Country and the City* (London: Chatto & Windus, 1973).

Wittgenstein, Ludwig, *Philosophical Investigations*, trans. G. E. M. Anscombe, 2nd edn (Oxford: Basil Blackwell, 1958).

Wyatt, John, *Wordsworth and the Geologists: A Correlation of Influences* (Cambridge: CUP, 1995).

Yeo, Richard, 'Reading Encyclopedias: Science and the Organization of Knowledge in British Dictionaries of Arts and Sciences, 1730–1850', *Isis*, 82 (1991), 24–49.
____ *Defining Science: William Whewell, Natural Knowledge, and Public Debate in Early Victorian Britain* (Cambridge: CUP, 1993).
____ *Encyclopaedic Visions: Scientific Dictionaries and Enlightenment Culture* (Cambridge: CUP, 2001).

Index